Symbolism, Synesthesia, and Semiotics, Multidisciplinary Approach

Symbolism, Synesthesia, and Semiotics, Multidisciplinary Approach

BONAVENTURE BALLA

Copyright © 2012 by Bonaventure Balla.

ISBN: Softcover   978-1-4771-5543-1
          Ebook   978-1-4771-5544-8

All rights reserved. No part of this book may be reproduced or transmitted in any form or by any means, electronic or mechanical, including photocopying, recording, or by any information storage and retrieval system, without permission in writing from the copyright owner.

This book was printed in the United States of America.

To order additional copies of this book, contact:
Xlibris Corporation
1-888-795-4274
www.Xlibris.com
Orders@Xlibris.com

TABLE OF CONTENTS

Dedication 7

Foreword 9

Author's Biography 11

Introduction 13

Chapter I: Symbol and Symbolism: Definition, Semantic and Philosophical Background 19

Chapter II: Esthetic Canons of Symbolism 53

Chapter III: The Acme of Symbolism through Cratylism: Synesthesia, Semiotics as its Tool of Hermeneutic Investigation 75

Chapter IV: Interests, usefulness of symbolism, its epistemological rupture 143

Conclusion 201

Illustrations: Study of Symbolist Texts: **Questions** 208

Glossary 229

Study of Symbolist Texts: **Answers** 239

Bibliography 293

Index 297

SYMBOLISM

Synthesis of the abstract and the concrete,
You are weaving the material and spiritual.
Meaning becomes exponentially discrete
Built under the aegis of semiotic ritual.
Ontogenesis of endless semantics
Lingering the analysis of any verse,
Infuse materials for hermeneutics,
Sift a glossary of the universe
Molded by the praxis of Cratylism!

SYNESTHESIA (TO THE MUSES OF SYNESTHESIA)

Sensory perceptions nurtured by the brain
Yeast of the limbic system honed by the strain,
Never dry up my pristine anamnesis!
Enthralling conundrum of our genesis,
Symbiosis of science and the ethereal,
Techniques of the mind refined in the aerial,
Heal my poetry stranded in my fieldwork,
Enrapture scholars gauging my work!
Sleet viaticum, spark off my intellect,
Inspire my lyre, make it perfect
And let this book soar in the heavens!

SEMIOTICS

Science of signs heralding semantics,
Emblem of truth nurturing delight,
Mother of logics, son of linguistics,
Inspire us with your refined insight!
O cosmic decipher of graphic gnosis
Teach us the delicacy of exegesis
In a world steeped in decrypting noesis!
Collusion of hermeneutics and ethos,
Seek meaning encrypted in chaos!

BONAVENTURE BALLA

DEDICATION

THIS BOOK IS DEDICATED

TO . . .

MY DAD AND MOM WHO TOILED FOR MY EDUCATION,

AIME CESAIRE, THE ACERBIC AND WINGED LYRE,

LAST BUT NOT LEAST, ALL THOSE WHO NEVER BELIEVED IN ME.

FOREWORD

Dear reader,

Have you ever heard about symbols and sounds or music inherently associated with colors? Have you ever heard about people who always dream in color, see sounds or hear colors, odors, and who cannot dissociate days of week, months, numbers and letters from specific colors? This phenomenon is real and called 'synesthesia'. It can be literary, scientific, and/or cognitive. It is analyzed within the framework of symbolism and neuroscience. It takes place in the left hemisphere of the human brain and the neo-cortex. It is activated by the limbic system and the tangling of two or more synapses. In this book I aspire to reflect on this phenomenon under the auspices of symbolism and neuroscience. However, I will emphasize the literary aspect of synesthesia (synesthesia as a metaphor) while pondering on symbolism as a general trend along with its scientific and cognitive aspects. I choose not to analyze symbolism as a literary school because I do not wish to get involved in heuristic and scholarly complexities that had featured symbolism as a literary school from the middle of the nineteenth century up to the end of this century. These complexities, albeit interesting, remain scholarly fruitless, and have historically reached their apex in the emergence of parochialism disseminated within the creation of ephemeral micro-schools, each of them claiming to have a monopoly of the symbolist truth: "Instrumentistes", "Vers Libristes", Decadent School, and "L'Ecole Romane". Accordingly, I will focus my analysis on symbolism as a trend, that is: on its major aspects and predominantly its origin, historical background, poetic idealism (culminating into synesthesia), esthetic canons, functions and, last but not least, its kinship with semiotics, psycholinguistics, and the anatomy and physiology of the human brain (neuroscience). Consequently, this book will give us a unique opportunity to engage an interesting and fruitful discussion on a holistic topic pertaining to literature, semiotics, science, psycholinguistics, and philosophy, to mention but a few.

AUTHOR'S BIOGRAPHY

Bonaventure Balla-Omgba is an Assistant Professor at Winston-Salem State University in North Carolina. He holds a PhD in French/Francophone Literature with a focus on Structuralist Semiotics, an MA in French, another MA in English and speaks seven languages: French, English, Spanish, Beti, Latin, Japanese and Fang. He has been a trilingual translator (English-French-Spanish) for 28 years and is currently a member of the American Translators Association (ATA). Dr. Balla has been a professor/instructor for 22 years in West Africa (13), France (1) and the US (8). He is the author of two collections of poems entitled *Astres et Désastres* (in French Language published in Belgium), and *Odes for Black Diamonds* (in English) published in the US. He has also written and published several conference papers and articles in English and French. He is finishing writing two books: *Euphoria and Dysphoria, and The Keys of Translation "Science."* Additionally, he is a black belt in Karate and has been practicing and teaching it for 27 years. He is an active member of several professional organizations.

INTRODUCTION

While the reader will be holding this book sitting in a cozy sofa, a naïve question will certainly cross his/her mind: why insert symbolism, synesthesia, and semiotics within the same framework? The answer is simple and coherent. Symbolism, synesthesia and semiotics, that I coin the three S (3s) for the purpose of my analysis, are organically linked. The first two are inherently connected in so far as synesthesia is one of the major esthetic canons of symbolism. As a matter of fact, symbolism and synesthesia are mutually inclusive. Both of them constitute a monolithic unit and officially goes as far back as Cratylus, Plato, idealist and spiritualist philosophers with the same doctrinal, esthetic, and linguistic implications. Both of them nurture the same aspiration: the finalization of absolute poetry and art. Thus, the study of symbolism presupposes that of synesthesia as well. Actually, symbolism reaches its peak through synesthesia with cratylism that is: the attempt to ward off the arbitrariness of the linguistic sign to systematize and highly refine its power. It follows that every artist or poet seriously affiliated with symbolism uses synesthesia or, at least, ponders on it significantly. However, to conduct a refined and cogent analysis on symbolism, one needs an effective tool of hermeneutic investigation. Semiotics, which can be conceived as the study of systems of signs to elucidate their meanings and show how they function in any graphic system for the purpose of communication, seems to be the ideal method to conduct such an investigation. Additionally, since semiotics examines signs and a combination of signs constitute a symbol (factually an exponential deployment of signs in a graphic system), it transitively analyzes symbols. Therefore, there is a deep and internal link between semiotics, symbols, and symbolism. From this standpoint, it appears that symbolism, synesthesia and semiotics are interconnected to form a triptych. In light of these considerations, they can be unified intrinsically and the study of these three S is not subject to sheer randomness. Consequently, in this book, I propose to study them by isolating their cardinal principles in order to highlight their origin, nature, goals, and pragmatic interests. In addition, given that, as far as I know, no researcher has ever provided an elaborate theory of literary synesthesia, this book aspires to fill this scholarly gap by providing readers, students, researchers in the field with a meticulous study of literary synesthesia. Accordingly, it is unique in the analysis of the theory of literary synesthesia. In it I also aspire to reflect on the philosophical, linguistic, and esthetic significance of synesthesia nowadays. As initially mentioned, I will utilize a semiotic approach with a neuroesthetic sub-approach, which surmises that my approach will also be buttressed by a multidisciplinary perspective and especially notions pertaining to literature, physics, psycholinguistics, neuroscience (a modicum of the anatomy and physiology of the human brain), and cognitive sciences in general. Neuroesthetics (also spelled 'neuroaesthetics') is a relatively recent sub-discipline of empirical aesthetics. Empirical aesthetics takes a scientific approach to the study of aesthetic perceptions of art and music. Neuroesthetics received its formal definition in 2002 as the scientific study of the neural bases for the contemplation, appreciation, and creation of a work of art. Neuroesthetics uses the techniques of neuroscience in order to explain and understand the aesthetic experiences at the neurological level. The topic attracts scholars from many disciplines including neuroscientists, art historians, artists, writers, and psychologists.

In actuality, symbolism through synesthesia applies to several fields of human knowledge. No field can claim either a monopoly or the jurisdiction of this convivially shared intellectual territory. Synesthesia is a complex notion and phenomenon. That is why sciences by themselves cannot provide a clear and exhaustive explanation of it. Likewise, because of its complexity, literature, semiotics, psycholinguistics or psychology combined cannot convincingly explain it. Accordingly, a fruitful conjunction of several fields (neuroscience, psychology, psycholinguistics, literature, semiotics, etc...) is necessary to understand it with all its sophisticated ramifications. That is also my reason for using a neuroesthetic approach as one of the best ways to analyze a holistic topic synthesizing arts/esthetics and neuroscience. Symbolism through synesthesia might therefore be the unified tool that bridges the gap between several fields of human knowledge all the more as it nurtures or, at least, opens up a broad spectrum of cognitive possibilities.

Moreover, we live in a time when overspecialization has reached its peak. At a certain point of human history overspecialization was necessary but, nowadays, if we need to survive as scholars and human beings in a more and more challenging world, we must think across disciplines, not within the narrow prism of our specialties because life and reality become more complex. To comprehend them and succeed in living in plenitude, it is high time for us to start thinking holistically. Thinking in this way will not only enhance the unity existing in human culture but, also and more importantly, endow us with the ability to decrypt the mysteries of the cosmos and govern it. As Dr. Murray Gell-Mann, Nobel Prize laureate for physics cogently outlines in his book *The Quark and the Jaguar*:

> What has always impressed me is the unity of human culture, with science being an important part. Even the distinction between nature and culture is not a sharp one; we human beings need to remember that we are part of nature. Specialization, although a necessary feature of our civilization, needs to be supplemented by integration of thinking across disciplines." (12)

It follows that, using a modicum of math or physics to fully understand a notion or concept apparently not linked with them can provide a much broader understanding of it. Several areas of human knowledge (for instance physics, neurology and synesthetic metaphors) might not, at a first sight, or for a layman, have any connections whatsoever, but, their close scrutiny might reveal surprising links because there are hidden connections between many fields of human knowledge and elements of the universe. Precisely, the ultimate stage of symbolism is to reveal them through the theory of correspondences conceived by Immanuel Swedenborg, a famous Swedish philosopher and scientist of the eighteenth-century.

Swedenborg inspired many symbolists and especially Baudelaire (who wrote several poems on the theory of correspondences, one of which was precisely entitled "Correspondences"), Mallarmé, Rodenbach, Yeats, T.S. Eliot, among others. Cosmologists have also studied these correspondences and proven that they do exist. Indeed, analyses conducted within the framework of quantum mechanics for instance have corroborated their existence by means of the quantum non locality principle. This principle has shown that:

- Several elements can be in different locations at the same time and influence each other;
- Something happening in Point A can impact something else occurring in Point B, regardless of the distance between Point A and Point B, because most elements in the universe are interconnected at a microcosmic level (sub-atomic scale, quantum level) and at a macrocosmic level (cosmological scale, super-galactic level). That is what Dr. Michio Kaku, a renown theoretical physicist brilliantly demonstrates in his book *Hyperspace* by underscoring the hidden symmetries existing in nature. He says:

> Simplicity. Elegance. These are the qualities that have inspired some of the greatest artists to create their masterpieces, and they are precisely the same qualities that motivate scientists to search for the laws of nature. Like a work of art or a haunting poem, equations have a beauty and rhythm all their own [. . .] In some sense, the equations of physics are like the poems of nature. They are short and are organized according to some principle, and the most beautiful of them convey the hidden symmetries of nature (130).

In light of these considerations, I believe that thinking across disciplines is important nowadays and researchers of all fields should collaborate through a common core and advocate open-mindedness. Unfortunately, some colleagues are so biased that they are prone to systematically sweep away from "the academic carpet" anything they had not yet studied in textbooks or labs (or anything that seems uncanny but that is empirically not). They do not know that the truth does not always manifest itself under the exquisite smile and gorgeous face of Ms. Clarity. It can be concealed. That is precisely why bona fide scholars need to dig in the surface, let themselves drift by the golden wings of open-mindedness and Mother Imagination to extirpate the truth (that is: pull it up by the roots after discovering it—> from Latin ex = out, stirps, stirpis = root)). De facto, if we scratch our heads and made some diligent efforts, this beloved Mother will open the door to invention and discovery for us. Is it not what the great genius of contemporary times did? Indeed, Einstein strongly believed that: "imagination is more important than knowledge". Accordingly, when scholars run into an epistemological obstacle or something they have not yet studied in textbooks or labs, they should use their imagination instead of pouncing upon it and vigorously disqualifying it as "unscientific" without prior scrutiny. This inspires me the following poem:

IMAGINATION

In the pristine shrine of our heuristic drive,
May we drift in the wings of imagination
As we shrink the limits of intellection!
Gear fuelling a matchless creativity,
Inspiration of our noetic activity,
Nurture the celestial seeds of masterpieces
And graft them in the core of your edifices!
Teach us to perform gorgeous parturitions
In the midst of skepticism and its deletions!
O Mother bearing the seal of aspirations,
Never stop creation, condition it to thrive!

BONAVENTURE BALLA

As I mentioned earlier, interconnectedness nurtures many disciplines of human knowledge. Precisely, symbolism through synesthesia bears the seal of this interconnectedness because it entails the conjunction of fields such as: poetry, semiotics, linguistics, psychology, a modicum of physics, and neuroscience (particularly areas like: the left hemisphere, limbic system, psychic areas (five sensory modalities), synapses, neurons, thalamus, and angular gyrus.). Therefore, I pledge to use a multidisciplinary approach to conduct my analysis of symbolism and synesthesia. Since one of my fields of expertise is French symbolism, I will focus my research on it and French symbolists in general. However, whenever necessary, I will also refer to symbolists outside France (Yeats, T.S. Eliot, Rodenbach, etc . . .), but my corpus will predominantly be French symbolism. Consequently, my study of symbolism will be determined by the semiotic and *epistemological status ascribed to French symbolism. At the end of this book there is a glossary and words preceded by an asterix (*) are explained in it. At the end of each chapter, there is a short summary of the main points that I have analyzed in each chapter. This will provide students, readers, or researchers with the opportunity to refresh their memories. Let us therefore explore and share an exhilarating intellectual experience under the aegis of symbolism, synesthesia, and semiotics but enlightened by literature, physics, neuroesthetics, arts, therapeutics, and cognitive sciences combined.

CHAPTER I

DEFINITION OF THE SYMBOL AND SYMBOLISM

A—DEFINITION OF THE SYMBOL/ETYMOLOGY

The word "symbol" and its *paronymic derivative "symbolism" stem from the Greek "σύμβολον" (sumbolon) which means: emblem, sign, token, or mark. Regarding the Greek verb 'sumbolein', it means "to throw together, to cast jointly". It is composed of the prefix "συν" (sun): combination, junction, coalition; and the root "βολή" (bolein): to throw or cast. It also expresses the idea of "sign, ticket, or contract". The semantic origin of the symbol goes as far back as the mist of times. Eons of years ago, a man was traveling to a very distant country. He had to find full board and lodging. Fortunately, he met a very kind man who accepted to host him for his entire journey. He was granted decent accommodation, nice food and everything he needed for his stay. Before taking leave of his host, he pledged to give him a present. Then, he took a jar and threw it on the ground. It was broken into two pieces. He cast both pieces on the table, took one and gave the other to his host, as a token of gratitude. He said: "From now onwards, you and I will be connected for ever. These pieces will seal our special link and represent our eternal friendship." This was purportedly one of the origins of the word "symbol". In this story, we see how the etymological meaning of the word functions (the pieces thrown together—> "sumbolein" in Greek = to throw together) and the connection set up between two realities, one standing for the other: a piece of jug (the concrete reality) representing friendship (the abstract reality). Another early—but yet to be proven—attestation of the word 'symbol' might come from Homer and especially in the *Homeric Hymn to Hermes* where Hermes on seeing the tortoise exclaims σύμβολον ἤδη μοι μέγ' ὀνήσιμον "*symbolon* of joy to me!" before turning it into a lyre. Presumably, here, the lyre stands for a concrete representation of the power of inspiration and the muses. In light of this cursory etymological inquiry, it appears that the symbol itself presupposes two elements or items thrown together (sumbolein), one standing for the other. One is usually concrete whereas the other is abstract. Therefore, a symbol is the concrete representation of an abstract reality. It is possible to find a cornucopia of symbols in the universe. For instance, post-Katrina New Orleans in the US (concrete reality) is considered as a symbol of resilience and determination/ sense of community (abstract). A pair of **scales** (concrete representation) is the symbol of **justice** (abstract reality). Likewise, a **dove** (concrete) is the symbol of **peace** (abstract). A **flower** (concrete), in general and a **rose** in particular, symbolizes **beauty or love** (abstract). A **lion** (concrete) is the symbol **of power, majesty, or courage** (abstract). A bird and its **wings** (concrete) symbolize **spirituality** or **the transcendence** of the soul over matter (abstract). A **crown** or a **scepter** (concrete) is the symbol of **kingship** or **power** (abstract). A **snake biting its tail inside a circle** (concrete) symbolizes a **palingenetic vision of the universe** (abstract), which means a vision positing that life is nothing but an endless succession of cycles, an eternal repetition of the same cosmic phenomena, or just a representation of eternity, endlessness, and the infinite **(From the Greek "palingenesis"-> "palin": again, and "genesis": birth)**. In Yeats's *Second Coming* the image of the 'gyre' (the last word of the first line) might fit in the palingenetic vision of the universe.

However, symbols are not limited to a pure connection between concrete objects and abstract realities because actions can also be endowed with an incontrovertible symbolic aspect. For example, putting one's hands up symbolizes surrendering to the enemy or deciding to give up a fight. Greeting somebody by bending at an angle of ninety degrees is the symbol of deep respect, and the larger the angle is, the deeper the respect becomes. Sub-Saharan Africans, Japanese and Koreans, among others, are familiar with such greetings because they are one expression of their rich culture. On the contrary, spitting on the ground while facing a person is the symbol of open contempt. Symbolism goes far beyond words, combination of words, concepts, and gestures. De facto, colors, geometrical figures, certain sounds, odors, gait, smile, look, dance, pause, rhyme scheme, rhythm, sounds can nurture symbolism. Numbers can be fraught with symbolism as well. Such is the case of seven (7), three (3), or twelve (12) for instance, are regarded as sacred numbers with specific symbolic meanings in numerology. *Dictionary of Symbols* provide more information in this special area. A particular poet or dramatist can organize or arrange rhyme schemes, sounds, rhythm in such a deliberate way that he definitely endows them with symbolism. In this case, they need to be analyzed in light of the particular context in which they are employed. That is why symbols can sometimes be personal. Indeed, certain authors, critics may assign symbols particular meanings which can vary from one writer or critic to another. In Baudelaire for instance, the albatros (albatross) and the cygne (swan) symbolize the genius, the wise man marginalized by his society because he is so far ahead of his time that his contemporaries can understand neither his talent, nor the magnitude of his faculties. With respect to the "azur", it usually stands for the ethereal, the ideal world, or the transcendental. That is the symbolism conferred upon it in "L'Albatros", "Vie antérieure". However, in Mallarmé's poetry the symbolism of the azur does not generally go as far as the transcendental. That is the way in which it can be assessed in the poem "Tristesse d'été ". However, in poems: "Azur", "Soupir" and "Renouveau", its symbolism becomes close to that of Baudelaire's poems. In the same vein, in Rodenbach's *Bruges-La-Morte* the recurrent construction "Le Démon de l'Analogie" symbolizes the vertical correspondences between sensory perceptions, and emphasizes the similarities between the voice of the late beloved (standing for the ideal world) and that of Jane standing for the material world). Besides, "Le Démon de l'Analogie" and its symbolism nurture the * isotopy of Platonism in this poetic prose novel. This expression is also used allegorically whereas in Mallarmé's *Poésies* it suggests nothing but simple correspondences without any transcendental symbolism or aspiration. Moreover, in the former, symbolism is reflexive, it suggests itself because it contains a technical device called "mise en abyme", that is: internal reduplication, systematized by the presence of mirrors, schemes of repetition or recurrent images designed to copy or imitate specific realities or notions. In the latter, it is not the case. Last but not least, in *Bruges-La-Morte*, Rodenbach makes a very interesting use of symbolism by blending poetry and prose, which leads to an original and eclectic approach of symbolism. On the contrary, in Mallarmé, this technical modus operandi does not exist. It follows that we can definitely talk about "meta-symbolism" in Rodenbach. The Greek prefix "meta" means "beyond, higher, over". Thus, meta-symbolism is a symbolism designed to express or suggest another form of symbolism (higher symbolism) to superimpose and formalize it. Meta-symbolism is reflexive. It expresses a more vivid aspiration to the ideal and, accordingly, a radicalization of Platonism or idealism. It is radicalized in Rodenbach and Baudelaire and less in Mallarmé. Therefore, symbols (and symbolism) slightly vary from Baudelaire to Mallarmé and from Mallarmé to Rodenbach in spite of their common affiliation with symbolism as a literary trend. All these factors show us that, apart from those that are universal and natural, symbols in general are not semantically static. They need to be contextualized, historicized because their meaning(s), layers of meaning, semantic implications, ramifications and perception are likely to change in space and time, and from one author to another. These semantic parameters also depend on a number of supplemental factors:

- The creative postulates of the writer using them;
- The heuristic postulates of the critic or theoretician studying them;
- The goals that the critic or theoretician wants to achieve in his/her research;
- The goals that the writer plans to ascribe to his/her work and its internal or/and overall organization.

Based on the aforementioned considerations, we become cognizant of the fact that the nature, function of symbols and symbolism can considerably vary in literature and arts. *The Princeton University Handbook of Poetic Terms* pertinently endorses this viewpoint. It states:

> The differences of opinion which exist today regarding the nature and function of symbolism in literature are due principally to the variety of ways in which the term is used in the service of different critical theories. This is true for many other terms as well, for a critic's use of any given term is

governed by the assumptions he makes about literature and the kind of knowledge he is interested in obtaining. One cannot therefore compare and contrast the interpretations of different critics without realizing what their assumptions and goals are and, consequently, how they use various terms.... [...] Thus, if symbolism refers generally to the use of one thing to stand for another, then its specific meanings will vary according to the framework in which this relationship is viewed. A symbol is a device of the poetic art when it refers to something in the poem as standing for something else in the poem; it is a power of poetic language when it refers to the way words and rhythms can evoke mystery, it is a function of the whole poem when it refers to the kind of meanings a literary work can stand for; it is a form of therapeutic disguise when it refers to the ways in which a poem stands for the working out of the author's inner disturbances; and it is an index of cultural values when it refers to the ways in which man's products reveal his attitudes. Since the word is thus capable of such protean meanings—some of the them overlapping at certain points—it is obvious best, when using the term, to specify the exact sense intended (275).

Accordingly, when analyzing symbols (especially those that are neither universal nor natural) one should be meticulous, flexible and open minded and invariably endeavors to be enlightened by a few *hermeneutic tools such as: socio-historical contexts, semantic field, goals the author wants to achieve, research or creative premises, ideological background of the author, etc . . . On the basis of all these considerations it dawns upon us that the range of symbols can significantly be stretched. Besides, it is important to know the conceptual framework and function of a symbol to understand the mechanism of its meaning(s), its implication(s), and ramifications.

The conceptual frameworks of symbols

Simply put, symbols can adhere to three distinctive conceptual frameworks: idiosyncratic or private, transcendental, and holistic. There is no absolute demarcation line between them because each form of symbolism can espouse one or more of them. So, from one symbolist to another there may be slight differences in terms of conceptual framework.

Idiosyncratic framework

Personal or private symbols, those **created** by the poet, belong to this framework. It is composed of symbols **created** or **recreated** by the poet out of an outstanding research on words, their semantic field and effects. Regnier's, Valery's and Mallarmé's use of symbols epitomize the paradigm of this framework. With Mallarmé for instance, words are meticulously chosen for their suggestive effects, evocative power or obsoleteness (mandora), rareness (Idumea, nixe) or sibylline aspect (ptyx, nixe). As Lagarde and Michard mention in their anthology *XIXeme Siecle*, "With Mallarmé, words are no longer algebraic signs. They become celestial signs." (531) Precisely, Mallarmé combines them to **create** symbols because a symbol is primarily a conjunction of signs. There are symbols: the fire, heaven, albatross, ice, lilies, soul, etc ... that each poet explores, develops, and creates. Where Baudelaire felt life grievously, and developed a style to express that experience, Mallarmé started with words and turned them into beautiful creations that evaded the exterior world. What mattered in Mallarmé's and Baudelaire's paradigm was the coherence of that inner vision, and the sheer beauty of the verse.

Transcendental framework

Here, symbols are not **created** but **discovered** and generated by the Neo-Platonist tradition which considers that poetry transcends the universe of appearances and apprehends divine truth itself. Plato had used myths (*The Myth of the Cave, Timaeus*), images ('prisoners', 'prison', 'cave') and symbols ('body' 'grave') to express his ideas, and the Neo-Platonists added a good deal of their own, from Roman, Egypt and Middle Eastern mythology, alchemy and astrology. By dint of using these symbols, poets exploited what we now call "**archetypes**" and, by the same token, emphasized the metaphoric nature of language since the symbol is both the ultimate goal of the metaphor and its eminently refined paradigm. In *The Symbolism of Poetry* Yeats outlines this exceptional power of the symbol. His reflections are particularly significant: "metaphors are not profound enough to be moving, when they are not symbols, and when they are symbols they are the most perfect of all, because the most subtle, outside of pure sound, and through them one can the best find out what symbols are." (12) For Baudelaire in particular the symbol became the supreme reality whereas for Poe it was the supernatural beauty. The *Dictionary of Literary Terms and Literary Theory* gives an insightful explanation of this framework: "[. . .] Concrete images are used as symbols to represent a general or universal ideal world of which the real world is a shadow. Sir Thomas Browne, long before theories of symbolism were abundant, suggested the nature of this in his magnificent Neo-Platonic phrase: "The sun itself is the dark simulacrum, and light is the shadow of God." (886) Among symbolists who complied with this tradition there were Baudelaire (*Correspondences, Benédiction, Vie Antérieure*), Rodenbach (*Bruges-La-Morte*), J.K. Huysmans (*A Rebours*), W.B. Yeats (*The Second Coming, The Magi*), and T.S. Eliot (*Ash Wednesday, The Waste Land*).

Holistic framework

This framework is holistic because it synthesizes several approaches: quest, intuition and mysticism. Indeed, here, symbols are **decrypted/ discovered** out of research, sheer intuition and mysticism/spiritualism, three powerful hermeneutic tools. Symbolism subscribes to the principle that the poet is a seer, clairvoyant, or a special artist whose intuition enables him to decrypt universal analogies, correspondences between the material and the ethereal. That is the official and standard conception and status of the poet within the framework of symbolism as a philosophical doctrine, trend or poetic school based on Platonism. The most eminent scholars of French symbolism such as Claude Pichois, Jean Pommier, Lyod Austin have validated it. Therefore, in this framework, symbols are more **decrypted** than **created** through heuristic work, intuition/ inspiration, and mysticism concomitantly. This supposes that they naturally pre-exist but are not accessible to the layman because he cannot perceive the supra-reality, this "other world". However, Baudelaire believed that the other world (—and, by the same token, the understanding and decrypting of its symbols) was not attainable through religious faith or mysticism but "à travers la poésie". For him "Through poetry the soul perceives "les splendeurs situées derrière le tombeau." *(Notes Nouvelles sur Edgar Poe, 7)*. We personally believe that poetry by itself does not guaranty the achievement of this goal because it needs to be supplemented by the quest for "poetic truth", that is, the systematic research focused on all the linguistic and semiotic arsenal that enables poetry to reach its most highly refined status: absolute poetry. In addition to this quest, the symbolist poet needs inspiration, intuition, and mysticism. Consequently, quest for "poetic truth", intuition and mysticism are the triptych leading to the holistic framework, which is the cornerstone granting the poet the opportunity to decrypt symbols. The holistic framework is very ambitious, but more appropriate to symbolism because it aspires to absolute poetry. Among symbolists who highlight this framework there are Baudelaire, Rodenbach, Yeats, and T.S. Eliot. Consequently, these three frameworks examine how the symbol is conceived by symbolism, and from one symbolist to another. How does the symbol function?

The function of the symbol

The way a symbol functions determines its meaning(s) and semiotic process, that is, the mechanism through which meaning unfolds and can be scrutinized within a system of signs for the sake of communication. Indeed, the symbol abides by the semiotic principle of semantic indirection as it is outlined by Michael Riffaterre in *Semiotics of Poetry*. According to it, human language can be misleading because words, constructions do not always mean what they say. They can say something whereas they mean something else and thus generate a semantic evasion process. The concept "semantic evasion" designates a semiotic device underscoring the meaning of a text "escaping" from its matrix or nucleus (point where it is generated) towards an unpredictable direction. The Latin word "evasum" expresses the idea of going out (ex = out, evado, asi, asum = go out, come out, escape), escaping, turning out towards a different direction. It follows that what the text **literally** says (predictable) is somehow contradicted by what it **actually** means and suggests (unpredictable). As such, it develops a surprise effect. That is exactly how a symbol works and, by virtue of that, it may be assimilated to a trope. "Trope" stems from the Greek 'tropos'" meaning "return, turn over, capsize". This etymology instructs us on the fact that the trope designates all stylistic devices in which words, constructions almost often mean the contrary of what they say. That is the case of metaphor, metonymy, synecdoche, irony, catachresis, symbol, etc . . . Therefore, the symbol functionally subscribes to the mechanism of the trope: **semantic indirection**. *The Princeton Handbook of Poetry Terms* underlines the modus operandi of symbols with much sagacity:

> "This term (a symbol) in literary usage refers most specifically to a manner of representation in which what is shown (normally referring to something material) means, by virtue of association, something *more* or something *else* (normally referring to something immaterial). Thus a literary symbol unites an image (the analogy) and an idea of conception (the subject) which that image suggests or evokes—as for example the image of climbing a staircase (the difficulty involved in the effort to raise oneself) is used to suggest the idea of raising oneself spiritually or becoming purified" (T.S. Eliot's *Ash Wednesday)* (273).

In this explanation there are two key-ideas:

1. -the semantic indirection and
2. -the referral to something *more* or something *else*. Besides, the "what is shown (normally referring to something material)" just designates what we have called "a concrete representation" whereas the "something *more* or something *else* (normally referring to something immaterial)" just designates what we have called "an abstract reality". It is therefore possible to synthesize the symbol by creating its hermeneutic code as follows:

HERMENEUTIC CODE OF THE SYMBOL:

HC SYMB = Σ (CR + AR). SEMANTIC INDIRECTION. ADDITIONAL SEMES

Where:

- "HC SYMB" stands for Hermeneutic Code of the symbol;
- "Σ" stands for Summation, Association or Combination;
- "CR" stands for Concrete Reality
- "AR" stands for Abstract Reality
- "." stands for Multiplied by

- "Semantic Indirection" refers to the semantic evasion process by virtue of which one says something to mean something else
- "Additional semes" designates the function of the symbol to mean something more since a seme is a minimal distinctive unit of meaning

Example 1: Our society needs a dove if not it will continue to be involved in wars and disappear.

DOVE—> Σ (DOVE (concrete)/PEACE (abstract))

    SUMMATION or ASSOCIATION of DOVE and PEACE
    SEMANTIC INDIRECTION

DOVE (concrete) PEACE (abstract)

    SOCIETY

By virtue of the semantic indirection principle DOVE does not factually and literally refers to the dove (the bird) but to PEACE. Accordingly, a *semantic transfer occurs.

DOVE (concrete)	PEACE/HARMONY, STABILITY, LOVE (abstract realities) + additional semes
margin of semantic Indirection	
SOCIETY	

DOVE does not only refer to PEACE but also to other realities: HARMONY, STABILITY, LOVE and its *paradigmatic variants because love, stability and harmony are mutually inclusive. For instance, in a society where there is love, there are generally peace, harmony and stability as well. Therefore, this instance shows that the meaning of a symbol is open ended. Indeed, the context in which it is used, its **semantic field**, ***paradigmatic variants** and ***syntagmatic axis** exponentially and ceaselessly work to extend its meaning by generating additional meanings.

In conclusion, the symbol adheres to the concept of referential fallacy, which definitely coincides with semantic indirection. The referential fallacy is used by Riffaterre in *Semiotics of Poetry* to stress the fact that readers or even most critics systematically tend to connect the substance of their reading materials with reality. This can prove to be a mistake because the materials of a literary work do not necessarily have a link with reality. They can be used figuratively rather than literally. This vindicates their nostalgia for the referential.

It is noteworthy that allegory, symbols and parables often occur in symbolism Since they can overlap, they need to be examined meticulously.

a—Allegory and symbols

An allegory is a development of symbols extended in a text, a work of art, or a masterpiece. Sometimes, it designates personified abstractions, concepts and endowed with specific functions in a text. An allegory can also be the representation of an idea by an image. For instance, the image of a human skeleton armed with a scythe is the allegory of death. The meanings of an allegory just like those of the symbol are hidden. However, the allegory tends to be less explicit than a symbol. "Allegory" stems from Greek "allegoria", composed of "allos" signifying "other", "alien", "stranger" and "agoria": "speaking". This suggests that allegorical constructions are generally composed of two basic meanings: literal and hidden, other ("allos"). However, an allegory can have more than two levels of meaning. The meaning (or meanings) is hidden, indirect, evasive, less explicit, and plural. The *Dictionary of*

Literary Terms and Literary Theory provides the following definition: "As a rule an allegory is a story in verse or prose with a double meaning: a surface meaning; and a secondary meaning. It is a story therefore that can be read, understood and interpreted at two levels (in some cases at three or four levels)." (21)

In short, an allegory says something and means something else because its reference seems to escape exponentially and ceaselessly. It therefore creates semantic indirection, that is a shift from mimesis to rhetoricity, which means: from what one says or represents in a text, be it oral or graphic (mimesis) to what one truly means or suggests (rhetoricity). Hence the necessity of deploying an intellectual effort to decipher an allegory. The effort is worth it because an allegory just like a symbol does have a didactic function. Indeed, it is often designed to teach us lessons usually encrypted subtly. For instance, Plato's *Allegory of the Cave* is precisely a tacit explanation of the universe of the essences (The spiritual/ideal world) reflected through the illusions of the material universe. As such, it postulates the refutation of the material universe by emphasizing that the latter is, if not a deceptive, at least a pale and distorted copy of the former. Thus, it deconstructs dualism in favor of monism. Consequently, by means of this allegory, Plato underscores the fact that most human beings delude themselves into believing in the material universe and they take its illusions/appearances for facts. It is just a deformed aspect, an illusion of the ideal world, that of the archetypes. At a much deeper level of meaning, Plato's *Allegory of the Cave* tacitly suggests the world of the fourth dimension (the time or space-time continuum) and the way its ethereal realities factually appear through the deformed prism we have in this material world. Moreover, most allegories are myths, that is: cosmogonical tales, that is: focused upon a cosmogony, a system of ideas conceived to explain the origin of the universe (from Greek "cosmos" = universe and "gonos" = form, side, formation). The *Allegory of the Cave* is also termed *the Myth of the Cave* and it is specifically a cosmogonical tale since it theorizes on a philosophical and cryptic explanation of the genesis of the cosmos. An allegory can sometimes function as an allegory and symbol simultaneously, or/and indistinctively. One the most compelling instances is the famous poem titled "*The Albatross*" by Baudelaire. This text is construed as both an allegory and a symbol. De facto, on the one hand, the Albatross, the protagonist of the passage, symbolizes the poet or the genius and his plight in a world where he is neither understood nor validated, which is eloquently expressed in these considerations: "His gigantic wings prevent him from walking/soaring" ("Ses ailes de geant l'empechent de marcher" (*Les Fleurs du Mal*, 7). On the other hand, at the allegorical level, *The Albatross* is a development of several ideas or symbols/ concepts highlighting the cripple condition of the poet, the genius, or the wise man in his society. He is so tremendously advanced and talented that his own people do not have the ability to understand the magnitude of his invaluable knowledge or erudition. He seems to live in another planet. To paraphrase Lamartine's words, we can say that he becomes like a fallen god who still remembers the heavens ("un dieu dechu qui se souvient des cieux" (**Lamartine,** *Méditations Poétiques,* **72).** In the same vein, let us just recall the predicament of Copernicus, Galileo, and Baudelaire himself. It took the society of their time more than one hundred years to fully appreciate their messages, the depth, and the magnitude of their knowledge and talents. Consequently, when a text meticulously and cryptically develops ideas or images designed to outline facts for didactic or philosophical purposes, it becomes an allegory. Most allegories are philosophical (*The Myth of the Cave*) and/or poetic. Regarding symbols, they are of three major types and each type is contingent upon their substance, semantic implications and functions. It is noteworthy that symbols, allegory can easily be mistaken as well as symbols and parables. It is therefore important to define them explicitly.

b—symbols, parables and allegory
Parables and fables, symbolism to allegory?

Parables and fables are easily confused with one another. Symbolism and allegory are similarly confused as well. In common parlance, a parable is a story or short narrative designed to reveal allegorically some religious principle, moral lesson, psychological reality, or general truth (aphoristic aspect). Rather than using abstract discussion, a parable always teaches by comparison with real or literal occurrences—especially "homey" everyday occurrences to which a wide number of people can relate. The word *parable* comes from the Greek term parabol! ("para'" means "beside," plus "bol" ("bolein": to cast, to throw), which means "a casting, putting, throwing, turning"), which the Romans called *parabola* in classical rhetoric. The etymology emphasizes the notion of practicality, and simplicity, and proximity (cast aside, close) as opposed to complex abstraction or sophistication. Well-known examples of parables include those found in the synoptic Gospels, such as "The Prodigal Son" and "The Good Samaritan." In some Gospel versions, the parables are announced with the phrase, "The Kingdom of God is like . . . ." Technically speaking, biblical "parables" were originally examples of a Hebrew *genre* called *meshalim* (singular *mashal*), a word lacking a counter-part in Greek, Latin or English. *Meshalim* in Hebrew refers to "mysterious speech," i.e. spiritual riddles or enigmas the speaker uses in story-form. It is only in the Greek New Testament that these *meshalim* are conflated with allegorical readings. Non-religious works may serve as parables as well. For example, Melville's *Billy Budd* demonstrates that absolute good—such as the impressionable, naïve young sailor—may not co-exist with absolute evil—the villain Claggart.

A fable is also a brief story illustrating a moral. Unlike the parables, fables often include talking animals or animated objects as the principal characters. The interaction of these animals or inanimate things becomes *aphoristic, that is, reveals general truths about human nature, i.e., a person can learn practical lessons from the fictional antics in a fable. However, the lesson learned is not allegorical. Each animal is not necessarily a symbol for something else. Instead, the reader learns the lesson as an *exemplum*—an example of what one should or should not do. The sixth century (BC) Greek writer Aesop is most famous as an author of fables, but Phaedrus and Babrius in the first century (AD) expanded on his works. A famous collection of Indian fables was the Sanskrit *Bidpai* (circa 300 AD) and, in the medieval period, Marie de France (1200 AD) composed 102 fables in verse. After the 1600s, fables increasingly became common as a form of children's literature. Since the distinction between symbol and allegory is very important, let us give additional explanations on them.

Symbolism and allegory (the following is a comparative and contrastive analysis of these techniques from online

Literary Terms, symbolism and Allegory)

An allegory involves using many interconnected symbols or allegorical figures in such a way that in nearly every element of the narrative there is a meaning beyond the literal level, i.e., everything in the narrative is a symbol that relates to other symbols within the story. The allegorical story, poem, or play can be read either literally or as a symbolic statement about a political, spiritual, or psychological truth. The word "allegory" derives from the Greek a*llegoria* ("speaking otherwise"): The term loosely describes any story in verse or prose that has a double meaning. This narrative acts as an extended metaphor in which the plot or events reveal a meaning beyond what occurs in the text, creating a moral, spiritual, or even political meaning. The act of interpreting a story as if each object in it had an allegorical meaning is called *allegoresis*. If we wish to be more exact, an allegory is an act of interpretation—a way of understanding—rather than a *genre* in and of itself. Poems, novels, or plays can all be allegorical. These can be as short as a single sentence or as long as a ten-volume book. The label "allegory" comes from an interaction between symbols that creates a coherent meaning beyond that of the literal level of interpretation. Probably the most famous allegory in English literature is John Bunyan's *Pilgrim's Progress* (1678), in which the hero Everyman flees the City of Destruction and travels through the Valley of the Shadow of Death, Vanity Fair, Doubting Castle, and finally arrives at the Celestial City. The entire narrative represents the average human soul's pilgrimage through temptation and doubt to reach salvation in heaven. Other important allegorical works include mythological allegories like Apuleius' tale of Cupid and Psyche in *The Golden Ass* and Prudentius' *Psychomachia*. More recent, non-mythological allegories include Swift's *Gulliver's Travels*, Butler's *Erewhon*, and George Orwell's *Animal Farm*. Now, for the sake of explicitness, it is necessary to set up a classification of symbols.

TYPOLOGY OF SYMBOLS

There are three major types of symbols: natural symbols, artificial symbols and philosophical symbols. It is necessary for some symbols to be historicized or contextualized because their meanings can vary in space and time. Therefore, assigning them specific functions or meanings is contingent upon societies, historical periods of time, or social contexts within the framework of which they were formalized, standardized or just created. Moreover, there are other types of symbols: mystical Personal, etc . . . Mystical symbols have a philosophical background and, in this respect, most of them can fall under the category of philosophical symbols. Personal symbols are those inherent in particular writers and persons. Some writers just like persons can have their own systems of symbols. That is why they can assign a number of concepts, ideas, systems of signs a specific form of symbolism likely to be comprehended in the particular context in which they are employed. These symbols comply with their vision of the world, literary sensitivity, affect, ideological convictions, particular exploitation of language. For the sake of classification, let us arrogate ourselves the right of naming them "idiolectic symbols". For instance, James Joyce, Baudelaire, to mention but a few, have in addition to natural, artificial and philosophical symbols, their own systems of symbols, idiolectic symbols (For instance "the azur "in Baudelaire is idiolectic. It symbolizes the ethereal). Therefore, to be understood, their idiolectic symbols should be analyzed, thought in light of the literary framework within which they are employed. It follows that there are definitely more than three types of symbols but since we cannot deal with all the symbols of the universe in this book, we have chosen to reflect upon the most important ones: natural, artificial, and philosophical.

1—NATURAL SYMBOLS

Natural symbols are those generated from nature. They are generally universal and identical in every society and culture. For instance smoke is the symbol of fire. **smoke => fire—>** Where there is smoke, it generally suggests that there might be some fire there as well regardless of the city, the country, the continent where I happen to be. It follows that smoke and fire are "monolithically" linked since smoke does not often exist on its own. It needs fire a posteriori to manifest itself. Likewise, water is a symbol of life, breath or air is a symbol of life as well. Without water, one cannot live. The case of breath, or air is much more compelling because it is absolutely impossible to live without air. If an individual is deprived of it in a few minutes, he/she will die due to **a** state of anaerobia, but one can still live a few days deprived of water and dies subsequently. That is why in several languages the word "breath" suggests the same idea and factually bears the same meaning. Indeed, the words "pneuma" in Greek, "spiritus" and "anima" in Latin, "Ki" in Japanese, "prana" in Sanskrit, "souffle" in French indistinctively mean: **soul, spirit, life, inspiration** (inhalation, penetration of the spirit, the soul, that is: life (In Latin "in" = in and "spiritus" = spirit, soul), **energy/ life-force, breath))**. All these words concurrently highlight the fact that inspiration is inherently connected to the notion of air, wind, breath and can be found at the very basis of creation and life. Therefore, air and breath are the natural symbols of life. It can be inferred that having life presupposes having air, being able to breathe, having energy and, accordingly, a soul. Consequently, air, as a symbol of life, is linked to it: **air => life**. Similarly, water as another symbol of life is linked to life: **water => life**.

The dove is another natural symbol. It stands for peace, love, seduction and beauty. In *Signs and Symbols, Sourcebook*, Nozedar highlights its qualities as follows:

> The dove carries a universal symbolism that is ubiquitous as the bird itself. The world over, it is associated with the feminine aspect, love, and peace. Pigeons and doves have carried messages for thousands of years and the notion of this particular bird as a messenger from the Gods is perhaps more pronounced than others. This is backed up by the story of the dove sent out by Noah to determine how far the Ark was from land; the bird returned with a sprig of olive branch, and the juxtaposition of the bird with the olive branch is a sign of redemption, peace, and resolution. Columba (Latin name)—the dove—is the "secret", covert bird for the United States, its soft reasonable femininity counterbalancing the masculine glory of the more visible and overt eagle. The dove is also the symbolic bird of Israel. The dove is a gentle-looking, rounded bird; its call is soft and seductive. It is often seen snuggled up to its partner, and it may be these characteristics that make it a symbol of the feminine, and love. (272)

The eagle: is considered the king of birds. It symbolizes majesty, power, charisma. It is the counterpart of the dove and the official symbol of the United States. In *Signs and Symbols*, Sourcebook Nozedar provides insightful comments on the eagle. She says:

> One of the most important archetypal bird symbols, the prominence of eagle is a worldwide phenomenon. The eagle is the "King of the Birds" and the "Lion of the Skies" and its use as a symbol is clear. It resembles power, authority, nobility, and truth, it is the ultimate solar symbol. In Greek the name

of the eagle ("aetos", not in the quote) shares the same stem as "aigle" (French for eagle, not in the quote either) meaning "ray of light (272, Ibid.)

These comments enable to understand why the United States, the greatest military and economic power on earth, has chosen the eagle as its official symbol.

The knee is a universal symbol as well. Nozedar pertinently explains **its symbolism**. She declares: "kneeling is a sign of subjugation, deference, and humility, and the knee itself is a symbol of power and strength since the joint supports almost the entire weight of the body. To "bring to the knee" is therefore symbolic of taking away power." (457, Ibid.)

2—ARTIFICIAL OR CONVENTIONAL SYMBOLS

Artificial symbols are those that are created conventionally by the society. Indeed, we live in a highly codified society where conventions had been generated for the sake of socializing, communicating, learning, teaching etc . . . Living in this society somehow coerces us to make symbols: artificial symbols also called conventional symbols. These symbols have been created by means of a common consensus between members of the society. Through such a consensus, people agreed to create a number of symbols to designate or suggest specific ideas, concepts, notions, values, life styles, and ideals. For instance, during the era of Hitler the Nazi decided to make the Iron Cross to represent the fascist ideology and system. The cross being a concrete representation of an abstract reality (fascist ideology). **The Iron Cross (or Cross Pattée or Eisernas Kreuz):** adopted as the Iron Cross in Prussia. During the First World War, it appeared on German fighter planes and tanks. Later, it became a fascist symbol in France, Portugal and other nations.

In addition, we can find symbols in many areas of human knowledge. That is the case of math **($2+2 \neq 9$ <)**, and road symbols and signs used for didactic purposes. To learn math, we need to learn symbols and how to decode them and understand their underlying principles. Likewise, to drive a car, we need to learn symbols, decode them and make driving a safe practice for us and everybody.

The symbols on the dollar bill

The dollar bill deploys a beautiful demonstration of symbolism. The symbolic elements seen on the dollar bill are predominantly conventional. Let us mention but a few: the Bald Eagle (official symbol of the United States); the 13 stars (symbol of the 13 states that joined the Union); the formula "E Pluribus Unum" (symbol of the unity/symbiosis stemming from plurality/diversity); the phrase "Annuit Coeptis", which is written around the top of the seal, translates as "[we] favor the things which have begun" and indicates that there is work yet to be done; and the banner around the bottom of the seal reads "Novus Ordo Seclorum" which means roughly "New order of the Ages". Nozedar adds this information regarding the dollar bill:

> On the reverse of the dollar the most noticeable emblem is that of the Great Seal of the United States. This takes the form of a pyramid with its cap severed and replaced by a triangle with an eye inside; this is the All Seeing Eye. The symbol has been associated with the Illuminati [. . .]

> Also featured on the dollar is the Bald Eagle, which is the official bird symbols of the United States. The eagle holds the olive branch of peace in its right talon, but it is hard to prepared to fight, too, as indicated the arrows in its claws. There are 13 fruits in the olive branch perfectly balanced by the 13 arrows. Another bird appears in the dollar, too, but it is hard to find. An owl—symbol of wisdom—is supposed to appear on the note. Above the eagle there is a crown of stars, again, 13 in total. This represents the number of the States that first joined the Union. The stars can be joined to create Solomon's seal, one of the most powerful of all symbols. That the stars combine to create another symbol is a clever nod to the phrase, which streams along on a banner underneath; E Pluribus Unum means "Out of many comes one" and refers to the many States that make our Union (340).

3—PHILOSOPHICAL SYMBOLS

Philosophical symbols are those that challenge our critical thinking/ analysis of life and existential realities. They postulate and spark a sharp reflection on ontological realities and urge us to find answers to *eschatological questions such as: why are we here on earth? What is the purpose of life? What is the nature of the soul? What is on the other side of life? What is the nature of reality? How does it function? Since symbolism does have deep philosophical foundations, it has the propensity of emphasizing the use of philosophical symbols. These symbols trigger a deep existential analysis and our mode of pondering on life and the universe in general. Accordingly, they hone a deeper understanding of existential facts and life. They are factually designed to arouse us, keep us awaked rather than alienate our conscience or make it sleep. Sometimes, their meanings can be eminently complex or sophisticated and, at that point, require a great dose of analytical thinking. Philosophical symbols are teleological, that is, geared toward a purpose, an end. The Greek stem "teles" means: "purpose, end". Thus, teleological relates to something that has an end, a specific purpose. One of the commonest philosophical symbols is the yin and the yang. The yin and yang are a symbol of the harmony between opposites. They tend to deconstruct the notion of duality or dualism because, at a higher level of reflection and understanding of cosmic reality, we become cognizant that duality turns out to be illusory. Dualism—((from Latin "duo" = two), as a philosophical system accrediting the existence and mechanism of two major opposite principles in the universe, is a distorted representation of monism ((from 'monos" in Greek = one single), philosophical system accrediting the existence and mechanism of only one major principle in the universe. For instance, the positive and negative seem to be opposed but actually they are nothing but complementary. One of the best illustrations of this principle is the two polarities of a battery. Their co-existence or co-presence generates electricity and light, one without the other will not provide light. From a mere philosophical standpoint, they are not opposed but complementary, interdependent, and both of their actions are necessary to activate light. They are the concrete manifestation of the same energy existing in the universe but expressed differently in terms of polarity (negative electricity and positive electricity). Similarly, the notion of male and female sounds to formalize a principle of opposition but they are complementary as well because both of them subsume negative and vice versa. Male anatomically subsumes a part of female and vice versa. The best illustration of this principle is expounded by Neo-Platonism within the framework of the concept of the androgynus, conceived as the harmonious, complementary existence of man and woman. Likewise, strictly speaking, light is not the opposite of darkness because within light one can find darkness (shade and shadow) and within darkness one can find light (reflection of light). Therefore, everything existing in the universe bears its complementary expression (not opposite stricto sensu) and extension within its genetic code. The yin and yang symbolize this cosmic harmony and complementariness. The concept regarding the harmony of opposites was substantiated and validated by Neo-Platonism and its followers especially Ammonius Saccas, Plotinus and Porphyry through the image and concept of the androgynus. It is appropriate to consider elements of the universe more in terms of complementariness than opposite. For instance, art is not really opposed to science(s). I strongly believe that true art is very complex and sophisticated. Its understanding requires a scientific approach or, at least, is conditioned by scientific principles. From this perspective, it is nothing but a different manifestation of scientific knowledge.

To put it mathematically, we can say that, art, just like science, is a different unknown of the same equation whose systematic solution will provide us with a full/exhaustive understanding of the universe so that we decrypt all its arcana and master it. This explains why great minds of the past were both outstanding scientists and artists (and philosophers): Aristotle, Plato, Leonardo Da Vinci, Pascal, Descartes, Leibniz, etc . . . De facto, there was not any clear demarcation line between science and arts because both areas were different manifestations of the same and indivisible reality. Consequently, art is not the opposite of science but its complementary. On the basis of these considerations, it is clear that the philosophical and semantic implications of the yin and yang can be extended to a wide range of fields of expertise and existential facts. From that perspective, they can help us perceive reality not in terms of classical dichotomies but as a refined monolithic whole.

ILLUSTRATION OF THE YIN AND THE YANG

In the following illustration, we clearly see that the deconstruction of dualism/ duality is a tangible demonstration that reality actually unfolds in monism rather dualism. As a matter of fact, each portion subsumes a particle of the other. Within the yin there is the yang (white dot). Conversely, within the yang there is the yin (black dot).

The Tao (can be conceived as the order of the universe): An ancient Chinese symbol used originally to represent a widespread belief in unity, polarity, holism, and magic

The tree is another philosophical symbol. The tree is composed of three major parts: the roots, the trunk, and the branches. The roots are in direct connection with the soil or the ground and all its components and sub-components. The trunk is in connection with the air. Regarding the branches, they are oriented toward the sky, the heavens. This three-fold connection suggests three distinctive universes: the elemental, the material, and the ethereal. The roots—through the sap—feed the tree and contribute to its growth. When it grows, it becomes the trunk, that is, an extension of the roots and the branches express the aspiration to excel and reach their peak. In light of these considerations, the tree, in its triptych configuration, symbolizes the aspiration for beings to transcendence by a gradual evolution from the elemental stage (roots), the material (the trunk), and the celestial (branches). The transcendence itself symbolizes the merging into the Absolute. Consequently, the tree symbolizes the merging of trinity (not in a Christian connotation here) into unity, the gradual and quasi-dialectical evolution of beings from the elemental (less refined, inferior = the roots/ infant/childhood) to the divine (more refined, Superior = the branches/merging into the Celestial or Deity). The vertical/erected position of the tree suggests the correctness, the appropriate attitude, and the moral rectitude one should adopt to finalize this ontological progress in order to reach the Absolute/ the Divine. As such, the tree suggests the progress from the lower stage of life to the sublime and, accordingly, from essence to existence (existence conditions self-awareness and, ultimately, transcendence), immanence to full transcendence. It is noteworthy that in the Kabbalah tradition there is a concept called "sephirotic tree", also known as "the tree of life", composed of specific stages each of which has specific esoteric meanings and corresponding to specific cosmic entities (angels, archangels, seraphines, dominions, gods, etc . . .). Each stage has a specific function in relation to every aspect of human beings: physical, physiological, emotional, mental, spiritual, metaphysical, etc . . . A deep understanding of the function of the sephirotic tree, its underlying complex symbolism and the constant and responsible practice of special exercises relating to it are supposed to grant a sincere seeker of higher truths the exclusive privilege to be in connection with transcendental beings and lives his/her life in his/her plenitude. However, here, we cannot elaborate on the cosmic and mystical complexities of the sephirotic tree because it goes far beyond the goals and scope of this book.

From a quite different perspective, the tree symbolizes another reality. Indeed, when it has a very straight position, in psychoanalysis and—particularly with Freud and even Fanon—it symbolizes, in light of its erection-like stance, the phallus and/or the sexual power of the male organ in proportion with the height of the tree. Thus, the higher and straighter the tree is, the more power the image suggests and develops. In *Black Skins, White Masks*, Frantz Fanon subscribes to this phallic symbolism. Nozedar backed this symbolism in his book *Signs and Symbols Sourcebook*. She states: "Trees, towers, standing stones—all have phallic connotations as symbols of strength, support, and also as foundation of life." (459) Therefore, it dawns upon us that in the analysis of a symbol, one should constantly take into account a number of factors such as: its cultural, socio-historical context, hermeneutic postulates, its semantic ramifications and underlying implications. Since some writers create their own symbols, what we term *creative idiosyncrasy must also be taken into consideration. It is the faculty inherent in a particular writer to conceive and create his/her own symbols under the fiat of his/her artistic freedom, imagination, vision of the world, or ideological convictions. Besides, symbolism may also be viewed as a concept.

Symbolism as a concept

As a concept, symbolism designates the systematic use of symbols in a text or any graphic system. The symbol is not designed to describe, narrate, but to suggest because it posits that reality is complex, mysterious, evanescent, perennially changing and moving. Its complexity and dynamism were expressed by the Greek philosopher Heraclitus who pertinently declared: "Everything passes. One never bathes twice in the same river". Thus, reality is perpetually fluxing and fluctuating. If it were static, simple and clear, one would be in position to describe it more effectively and efficiently. However, it appears that it is not the case. Accordingly, one can just suggest it rather than describe it. Moreover, since it ceaselessly changes, it becomes more complex. The best way to present this complexity is to use the power of suggestion epitomized by the symbol with its ability to express the complexity and multifariousness of reality because suggestion is much stronger, subtler than description or narration. As a matter of fact, a symbol is semantically very rich. Its meaning exponentially and endlessly generates other meanings. That is why it subscribes to the hermeneutic non-exhaustiveness theory of the sign as it is stated by Gerard Genette in *Figures II*. Consequently, since its meaning ceaselessly escapes and produces other meanings, its rendition and analysis become a manipulation. Hence its adequacy to be used and exploited to suggest reality. It is also important to underscore that signs should not be confused with symbols. There is a clear distinction between these two notions. A symbol is factually composed of several signs. With respect to signs, in *Lecture Sémiotique* Debove provides a pertinent and cursory definition: a linguistic element having a signified, that is, a meaning and signifier, an expression. Signs combine to constitute a symbol." **(27)**

We have examined the origins and definition of the word "symbol". How about "symbolism"?

B—DEFINITION OF SYMBOLISM

"Symbolism" is composed of the root "symbol" and the suffix "ism". This suffix is very prolific in English and most neo-Latin languages. As a matter of fact, it has helped to create and designate a myriad of concepts, ideologies, doctrines, philosophies, systems of thought, ways of life, life styles, etc . . . In light of its etymology and the semantic hints of the suffix "ism", symbolism can be conceived as a system of thought, a philosophical doctrine, a poetic school, maximizing the use of symbols to highlight the rich subtlety, the complexity of existential realities and universe in general. As a literary trend/movement, it does maximize and standardize the use of symbols and suggest the highly refined relationships that take place between two polarities: materialism and idealism underlying the material world and the ideal world respectively. Since this book focuses on symbolism as a literary trend, we will deal with this perspective later on and substantially.

1—SYMBOLISM AS A PHILOSOPHICAL DOCTRINE AND SYSTEM OF THOUGHT

As a philosophical doctrine and theory, symbolism is in direct connection with Platonist Idealism. The fact of the matter is that, the roots of symbolism (at least officially) go as far as back as Plato and posits the existence of two worlds: the material and the ideal. The former mirrors the latter and is elucidated by idealism, a theory studied by Plato and his disciples and eminently exemplified in the allegory of the cave. This philosopher considers that the material world is nothing but a pale reflection of the transcendent world. The former is un-factual. Plato called it the world of appearances whereas the second is factual. He named it the world of essences or archetypes. Besides, the former is fickle, transitory, the latter is permanent, eternal. Plato had two masters: Heraclitus and Parmenides. Heraclitus believed that everything moves, passes, and fades away (material world) but, for Parmenides, there are eternal and perfect essences in an ideal world. Plotinus, a Neo-Platonist, and his disciples will call them "entelechies".

Platonism is the synthesis of Parmenides's and Heraclitus's systems and symbolism is based on Platonism. As a philosophical expression of Platonism, symbolism posits that there are two universes: the material and the supra-terrestrial. In symbolism, the role of the poet is to reveal the connections between these two worlds through symbols also called correspondences. However, in Platonism the same role is assigned to the philosopher. Therefore, the role assigned to the poet in symbolism is the same as that of the philosopher in Platonism. Plato formalized his theory of correspondences through his work the *Myth of the cave*

2—SYMBOLISM AS A POETIC SCHOOL AND LITERARY TREND

As a literary trend and poetic school, symbolism is expounded through its historical background and the function of the poet within symbolism.

BRIEF HISTORICAL BACKGROUND AND FUNCTION OF THE POET

a—BRIEF HISTORICAL BACKGROUND

As a literary trend, symbolism was heralded by Baudelaire and theoretically started at the end of the nineteenth century, circa 1870-1880. However, it will truly cover the second half of the nineteenth-century. Symbolism highlighted a new epistemological paradigm. Indeed, it reacted against classicism, neo-classicism, romanticism, realism, positivism and naturalism because these literary schools were construed as exhausted, "breathless", and, accordingly, incapable of expressing language, concepts and ideas optimally to sublimate reality and nature. Classicism and neo-classicism were too rigid and static. Thus, they could not grant the poet the artistic freedom necessary for the exercise of his creativity. Boileau's *Art Poetique*, conceived as a true manifesto of Classicism, prescribed very strict rules under the ferula of which creation became factually impossible or, at least, very difficult to "carry out". Regarding Romanticism, it deformed creative works because of excessive lyricism/emotional effusion. With respect to realism, symbolism posited that its fieldwork, reality, was illusory, fickle, evanescent, unstable and, therefore, unable to maximize the symbolist project and corpus. This literarily disqualified realism as a school likely to use reality to support symbolism and its ambitious project. Regarding positivism and naturalism, they stifled or shrank creativity with their intellectual straight jacket and arrogance. Indeed, positivism was close-minded and claimed to have a monopoly of truth since it arrogated itself the right of explaining and clinically verifying everything as if it were gratified with divine omniscience. Consequently, on the 'ashes' of all these literary schools, symbolism designed a brand new modus operandi and heuristic perception of poetry featuring the ideal of absolute poetry whose seal was marked in 1857, the date of the publication of Baudelaire's *Les Fleurs du Mal*. This seminal work highlighted Baudelaire's unique exploitation of the poetic material, his tremendous visionary power, sophisticated mastery of imagery and, in short, his pristine effort to materialize ideal poetry epitomized by a very powerful use of poetic language, metaphors, symbols and music in particular (basis of Cratylism). Such a linguistic asset was actually regarded as a paragon guiding symbolists in their aspiration to achieve absolute poetry. Symbolism drew its inspiration from a number of philosophical systems and especially: idealism, Platonism, and mysticism. From that particular standpoint, idealist philosophers such as: Plato, Plotinus and, later on, Kant, Fichte, Schelling, Hegel, Schopenhauer, Spinoza and Mallebranche can be regarded as the precursors or inspirers of symbolism. Most of these thinkers implicitly or explicitly fed the roots of the symbolist tree. Another thinker, the Swedish mystic and philosopher, Immanuel Swedenborg, brought a substantial contribution to symbolism through his famous theory of correspondences. He presented it in his book *Arcana Caelestia* and Baudelaire as well as several future symbolists—Yeats, T.S. Eliot, Rodenbach, Huysmans, etc . . . — certainly borrowed it from this book or at least from Swedenborg's mysticism or doctrine. As one of the real precursors of symbolism, Baudelaire informally set up the esthetic canons of this school. He and German philosophers of idealism and spiritualism can collegially be regarded as the "trunk of the symbolist tree". Baudelaire really brought this school a step further. In his famous sonnet titled "Correspondances" he became the very first poet still unofficially and informally affiliated with symbolism to use the word "symbols" and suggest its meaning tacitly: "La nature est un temple où de vivants piliers/ Laissent parfois sortir de confuses paroles/L'homme y passe à travers des forêts de **symboles** / Qui l'observent avec des regards familiers". Additionally, he suggested and adopted the use of one of the major esthetic canons of symbolism, synesthesia, bequeathed to him by Swedenborg. Synesthesia was also known as "horizontal correspondences".

They were magisterially exposed in his quatrain of the celebrated poem "Correspondances": "Comme de longs échos qui de loin se confondent/ Dans une ténébreuse et profonde unité/Vaste comme la nuit et comme la clarté/**Les parfums, les couleurs et les sons se répondent" (27)**

"Les parfums, les couleurs et les sons se répondent" refer to the combination of different sensory perceptions but so artfully attuned that they finally constitute a highly sophisticated organic unity precisely named "synesthesia" or "horizontal correspondences". Moreover, through the sonnet "Correspondances" Baudelaire suggested the function of the poet, which will consensually be the one that most symbolists will **adopt**. The following are an "organization-chart" and diagram of the symbolist tree highlighting the most representative inspirers, precursors and adepts of symbolism.

DIAGRAM OF THE SYMBOLIST TREE AS A TREND:

BRANCHES
  RODENBACH, HUYSMAN, MAETERLINCK,
  VILLIERS DE LISLE D'ADAM (Belgian symbolism)
  Mallarmé => PAUL VALERY
  YEATS, T. S. ELIOT (Anglo-Saxon symbolism)

TRUNK
  KANT (Idealism) => HEGEL, FICHTE, SCHELLING,
  SCHOPENHAUER (Post-Kantian Idealism)
  KIERKEGAARD, SPINOZA
  SWEDENBORG => proto-symbolists or precursors:
  BAUDELAIRE, NERVAL
  BAUDELAIRE => RIMBAUD, VERLAINE, Mallarmé
  GERMAIN NOUVEAU, TRISTAN CORBIERE,
  CHARLES CROS

ROOTS
PARMENIDES (idealism/essences)
HERACLITUS (materialism/appearances)
"Everything is static" "One never bathes twice in the same river, everything passes"
PLATO (Platonism) =>PLOTINUS (Neo-Platonism): SYNTHESIS OF PARMENIDES'S & HERACLITUS'S THESES

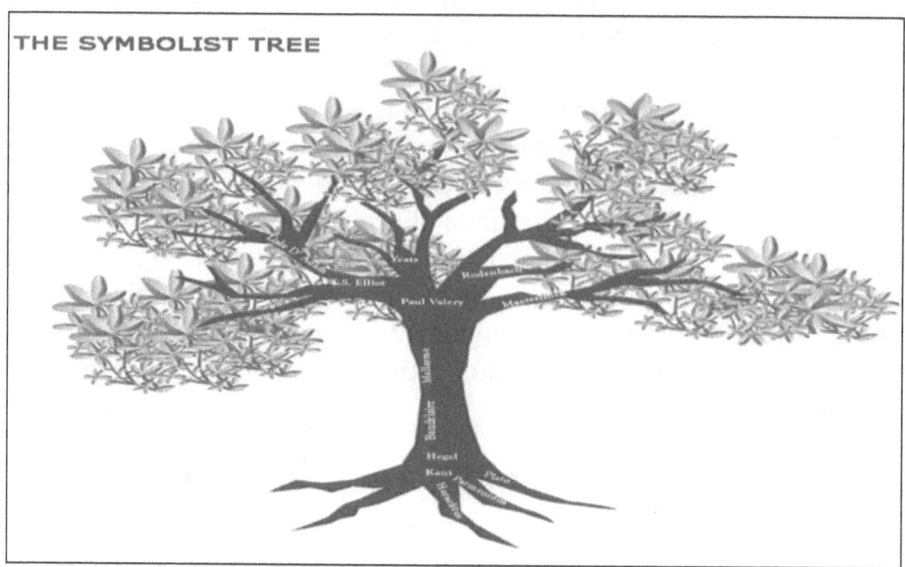

b—THE FUNCTION OF THE POET FROM A SYMBOLIST PERSPECTIVE

According to symbolists especially Baudelaire and his followers, the poet is an outstanding creature. He is an inspired, visionary, clairvoyant and beacon. In his poem entitled *Les Phares*, which means 'beacons' (52), Baudelaire emphasizes this special function by comparing artists (Rubens, Leonardo Da Vinci, Rembrandt, Michel-Ange, and Watteau) and poets to beacons, those privileged characters entrusted with the mission of guiding and leading the society. The Dictionary of Literary Terms and Literary Theory provides substantial information regarding this function: "Baudelaire and his followers created the image of the poet as a kind of seer or voyant, who could see through and beyond the real world to the world of ideal forms and essences. Thus, the task of the poet was to create this 'other world' by suggestion and symbolism; by transforming reality into a greater reality and more permanent reality." In the same vein, in *Notes sur Edgar Poe* (1857) Baudelaire emphasizes the seer faculty of the poet: "C'est cet admirable, cet immortel instinct du Beau qui nous fait considérer la Terre comme un aperçu, comme une correspondance du ciel. La soif insatiable de tout ce qui est au-delà, et que révèle la vie, est la preuve la plus évidente de notre immortalité. C'est à la fois par la poésie et à travers la poésie, par et à travers la musique que l'âme entrevoit les splendeurs situées derrière le tombeau". In light of this emphasis, we are prone to consider that the poet's elating task will consist in intuitively grasping these mysterious correspondences. Precisely, the poems "Correspondances" and "Elevation" among others corroborate the role of the poet as a seer endowed with the ability to perceive these ethereal splendors. Such a task stems from Platonism and is similarly enlightened by Platonism, a philosophical system positing that there are two worlds: the material and the ideal also known as the visible and the invisible world or the material world and that of the essences. The first is a pale, distorted and deformed reflection of the second. Hidden and subtle correspondences circulate between these two worlds. Unfortunately, most human beings have limited faculties. Accordingly, they can neither comprehend, nor see/ figure out these correspondences. Only the philosopher and the poet are endowed with the faculty of perceiving and decrypting them. In *Volume 7 of The Republic*, by means of an allegorical tale, Plato elaborates on the nature of the two universes and the condition of prisoners chained and kept in a cave. He acknowledges that our intuition and inspiration are limited and because of rationalism and skepticism we shut down or shrink our emotional intelligence, which radicalizes our metaphysical sub-condition by preventing us from perceiving, sensing or detecting the subtle correspondences between the ideal universe and the terrestrial universe. Let us give a synopsis of Plato's allegory here. In it Plato tells about prisoners chained in a cave. Nevertheless, a small hole located in it enables them to see the light coming from the outside world. Such a light also conveys images from this world. Still, these images are distorted, un-factual realities of the outside world. Unfortunately, because of their shrunk intuition, analytical myopia and flaws, the prisoners confuse the images with the very reality of the outside world. Consequently, they delude themselves into believing that what they see is mere facts. However, if by chance, one of them, through an incredible feat, manages to escape from the cave to reach the real world, he will find out an amazing universe. He will then realize that he and his former cave-mates were utterly wrong. Moreover, if he decides to revisit his cave-mates and tell them the true story, none of them will believe him. He is even likely to be marginalized. So, the allegory of the cave is by nature a didactic and symbolic tale of the material and the spiritual universe.

The cave is a symbol of the material world and the outside world that of the ethereal world. With respect to the prisoners, they symbolize human beings. Regarding, the prisoner who visited the real world and came back to tell his fantastic story, he can be equated with the philosopher or the poet. In light of all these considerations, we become cognizant of the fact that Plato's allegory is an accumulation/development of symbols to illustrate an existential and philosophical truth. That is why it is precisely known as an allegory (accumulation and development of symbols). Additionally, it is a myth, that is: a cosmogonical tale, (not an un-factual story, the myth conceived as an un-factual story is nothing but an erroneous or at most a very superficial definition given by certain dictionaries.) a narrative providing a system of ideas that give an explanation of the origin of the universe philosophically, symbolically, or/and poetically. In this respect, it is intrinsically didactic. Last, but not least, Plato's myth elucidates symbolism and the central positioning and the function of the poet within this literary school: a privileged authority, an inspired, a magus who can see hidden correspondences between the invisible and the visible. More importantly, he is also a decipherer of correspondences and universal analogies. By virtue of his faculties, he becomes the beacon for his society. Such is the brief historical background of symbolism. What are the major esthetic canons of symbolism or more specifically its theoretical program? That will precisely be the goal of chapter II.

SUMMARY OF CHAPTER I

The word "symbol" stems from Greek. Indeed, the prefix "συν" ("sun") means "together, combination, association" and the verb "βολή" ("bolein") means: to throw. In light of this etymological break down, It dawns upon us that the word "symbol" suggests the idea of "throwing together". The semantic origin of this word was lost in the mist of times. However, a number of accounts tell about a foreigner who was traveling to a far country. He was welcome by a munificent host. The latter provided him with a decent accommodation and delicious food. Thus, before taking leave of him, the traveler pledged to give him a present. He took a jug and threw it on the ground. It was broken into two pieces. He threw both, took one and gave the other to his host, as a token of gratitude. He said: "From now onwards, you and I will be connected for ever. These pieces will seal our special link and represent our eternal friendship." In this tale, each piece of jug is a concrete representation of an abstract reality, the friendship. It seems that this tale gave the key to the definition of the symbol: a concrete representation of an abstract reality. However, symbols are not limited to a pure connection between concrete objects and abstract realities because actions can also be endowed with an incontrovertible symbolic aspect. For instance, greeting somebody by bending at an angle of ninety degrees is the symbol of deep respect, and the larger the angle is, the deeper the respect becomes. Symbolism goes far beyond words, combination of words, concepts, and gestures. De facto, colors, geometrical figures, numbers, certain sounds, odors, gait, smile, dance, pause, rhyme scheme, rhythm, can nurture symbolism. A particular poet or author can organize or arrange rhyme schemes, sounds, rhythm in such a deliberate way that he definitely endows them with symbolism. That is why symbols can be personal. A symbol is polysemic and complies with the semiotic principle of semantic indirection. According to it, human language can be misleading because words/constructions do not always mean what they say. They can say something whereas they mean, suggest something else and thus generate a semantic evasion process induced by semantic indirection. The function of a symbol can be didactic, artistic, ideological, or philosophical. There is a cornucopia of symbols in the universe. However, most of them fall under three major categories: natural or universal symbols, artificial, and philosophical. symbols, and philosophical symbols. Natural symbols are those generated from nature. They are generally universal and identical in every society and culture. e.g. breath is the symbol of life. Its presence generally suggests that of life as well regardless of the people, the country, and the continent concerned. Artificial symbols are those that are created conventionally and consensually by the society. e. g. the eagle stands for majesty, nobility, charisma and power, is regarded as the lion of the skies. It was conventionally chosen to symbolize the United States. Philosophical symbols are those that spark critical thinking on ontological realities, existential phenomena, or our raison d'être on earth. e. g. the yin and yang is a symbol of the harmony between opposites, their co-existence and the deconstruction of duality and dualism. Apart from those that are universal and natural, symbols, in general, do not have a rigid meaning. They need to be contextualized because their meaning(s), layers of meaning, semantic ramifications can change in space and time. Their meanings also depend on the creative postulates of the writer using them; the heuristic postulates of the critic or theoretician studying them; the goals that the critic or theoretician wants to achieve in his/her research; the goals that the writer plans to ascribe to his work, its internal/ external, and overall organization.

CHAPTER II

ESTHETIC CANONS OF SYMBOLISM/ PROGRAM OF THE SYMBOLIST ESTHETIC THEORY

A better understanding of symbolism and its esthetic canons is pre-conditioned by that of the major goal of this literary school or trend. As a matter of fact, symbolism craves for materializing absolute poetry. The idea of absolute poetry or pure poetry focuses on the systematic exclusion of whatever is alien to poetry, which entails the theorization of a wide range of ambitious technical devices geared towards the achievement of this ideal. To comply with such an ideal, symbolism used a number of key-principles: the transcendence of reality to suggest hidden connections between the material and the immaterial universe (to reach this goal the poet will be ignited by inspiration, intuition, and mystery); the power of suggestion (use of symbols; systematization of connotation at the expense of denotation; use of a special, erudite lexicon (rare words, sophisticated erudition); the adoption of the collage, music, uneven verses, and free verses; and the use of synesthesia.

I—THE TRANSCENDENCE OF REALITY

Symbolism aspires to transcend reality to reach the ethereal world, that of the essences. The *Dictionary of Literary Terms and Literary Theory* provides pertinent explanations about this seminal process. It states: "Baudelaire and his followers created the image of the poet as a kind of seer (q.v.) or voyant, who could see through and beyond the real world to the world of ideal forms and essences[. . .] The attainment, in transcendental symbolism, of the vision of the essential Idea was to be achieved by a kind of blurring of reality so that the ideal becomes clearer" (887). Additionally, symbolism posits that reality, just like poetry, is enshrouded by mystery and hidden correspondences between the material and the ideal worlds. Indeed, for pure symbolists, poetry is conditioned by mystery, a kind of sacredness and one of the major functions of the poet is to unravel this mystery. For that reason, Mallarmé, the leading figure of symbolism, considered that poetry was an "archa firmata", which means a closed safe or close strongbox in medieval Latin. This metaphorical construction emphasized two ideas:

- Poetry is a treasure;
- The layman is neither intellectually and emotionally prepared to open it nor decipher its invaluable substance.

In 1862 he expounded these two ideas. He stated in *L'Artiste*: "Toute chose sacrée et qui veut demeurée sacrée s'enveloppe de mystère. Les religions se retranchent à l'abri des arcanes dévoilés au seul prédestiné: l'art a les siens . . . J'ai souvent demandé pourquoi ce caractère nécessaire a été refusé à un seul art, au plus grand. Je parle de la poésie . . ." ("Any holy thing and that aspires to remain so surrounds itself with mystery. Religions are secluded in the shield of arcana revealed only to seekers: art has its own . . . I have often wondered why this necessary aspect has been denied to one art, the greatest. I mean poetry . . ." It dawns upon us that symbolism was conceived in a background of mystery that fed its roots in metaphysics, Platonism and idealism because symbolism synthesizes these three philosophical systems. Precisely, metaphysicians such as Kant, Kierkegaard, Schopenhauer, Platonists and Neo-Platonists, have pondered upon the mystery enshrouding reality. This pondering reached its paroxysmal refinement with Swedenborg and was developed into the writing of *Heavenly Secrets*. Later on, Kant's *Foundations of the Metaphysics of Morals* endorsed the same theory. De facto, Kant acknowledged that we human beings do not know the **nomenon** that is: the mystery, the essences, or things in themselves. By 'nomenon' he meant a concept made up of three components: the soul, the cosmos, and God. His contention was that only the phenomena (that is the appearances) are accessible to us, not the nomena (essences) because our intellect cannot fully grasp their complex and higher truth. That is why the symbolist poet, through poetry, is involved in decrypting this mystery to reach the world of essences. For that purpose, he will use emotion, intuition and, more importantly, the power of suggestion instead of description. Actually, symbols, synesthesia are the technical devices that nurture such a power. Synesthesia, also called horizontal correspondences, turns out to be the elements **suggesting** the connection between reality and the ideal. Accordingly, whereas **description** fits in Parnassian, positivist, realistic, or naturalistic paradigms of pre-symbolist literature, **suggestion** fits in that of symbolist literature. The second paradigm is found in the most representative texts of symbolism: Baudelaire's *Correspondances*, Elevation, *Invitation au Voyage, Le Voyage, Harmonie du Soir, L'Albatros, Le Cygne*; Mallarmé's *Les Fenêtres, Apparition, Grand-Oeuvre*; Rimbaud's *Bateau Ivre, Illuminations, Une Saison en Enfer*; Verlaine's Fêtes Galantes, Jadis et Naguère, Rodenbach's *Bruges-La-Morte, Ciel Natal*, Yeats's Second Coming, Sailing to Byzantium, T.S. Eliot's *The Waste Land*. How does the paradigm of suggestion function?

II—THE POWER OF SUGGESTION

The power of suggestion is governed by a set of technical devices by virtue of which it functions: use of symbols/synesthesia; the quasi—systematization of connotation at the expense of denotation; use of a special, erudite lexicon (rare words, neologisms created out of Greek and Latin, obsolete words); the collage; use of free, loosen, and uneven verses; and music.

i—Use of symbols

Symbols were used extensively because of their semantic power and prolificacy. The symbol is semantically very rich because one cannot really exhaust its meaning. Its reading subscribes to the theory of hermeneutic inexhaustibility of the sign as it is described by Genette in *Figures II*. Genette pertinently explains this theory in these terms:

> La littérature suggère que le monde signifie, mais "sans dire quoi"; elle décrit des objets, des personnes, rapporte

> *des événements, et au lieu de leur imposer des significations certaines, figées, comme le fait la parole sociale (et aussi, bien sûr, la "mauvaise" l ittérature) elle leur laisse, ou plutôt, leur restitue, par une technique très subtile (et qui reste à étudier) d'évasion sémantique, ce sens tremble, ambigu, qui est leur vérité.* (90)

It is definitely this process of semantic evasion, this shaking, ambiguous meaning that points out the nature of the symbol. As a matter of fact, the symbol is designed to suggest and, as such, it becomes endowed with a substantial amount of meanings. It follows that it acts as if, through an ontogenetic process, it were targeting towards perennially generating language by the power of suggestion. In the long run, its meaning ceaselessly generates other forms of meaning and its interpretation becomes a highly sophisticated heuristic exercise driven by sheer manipulation. Therefore, suggestion by means of the symbol, has an apophantic function. Such a function consists in creating and especially creating meaning. As a result of this, the text becomes like a kind of *natura naturans, that is: a never ended nature, an open-ended system in which each element, units of meaning constantly contributes to generate additional units of meaning. Barthes terms these additional units of meaning "the obvie" in *L'Obvie et l'Obtus*. Accordingly, it appears that meanings become perennially "opened" to other meanings, suspended and they deny the possibility of producing an ultimate, transcendental signified. In this respect, Barthes states: "*Le poétique est très exactement la capacité symbolique d'une forme: cette capacité n'a de valeur que si elle permet à la forme de "partir" dans un grand nombre de directions et de manifester ainsi en puissance le cheminement infini du symbole, dont on ne peut jamais faire un signifié dernier et qui est en somme toujours le signifiant d'un autre significant* "(46)

ii—The quasi—systematization of connotation

One of the cardinal attributes of symbolism is epitomized in a diligent attempt to decrypt the mystery of the universe, to suggest the ideal, and the ineffable with a view to opening the doors of the sublime. However, since it is impossible to suggest the sublime in a vulgar language, symbolism will utilize a special vocabulary to reach such a goal. It will try to systematize connotation at the expense of denotation to purge poetry from linguistic prostitution and, so doing, "destroy common meaning", any word and anything likely to trivialize poetry. That way, it will infuse poetry with a real new life. In *Trésor des Grandes Oeuvres de la Littérature Française* Henri Mitterand examines the systematization of connotation in symbolist poets. Here is the core of his analysis:

> *Il faut vider le mot de son acception triviale pour le faire résonner, le faire étinceler en le nommant. Il y a la quelque terrorisme: la volonté de détruire le sens commun. Contre la dénotation Mallarmé préconise les connotations, contre la nomination, il met avant la suggestion, d'où l'axiome fameux: "nommer un objet c'est supprimer les trois quarts de la jouissance qui est de faire deviner peu à peu (70)*

Within the framework of this linguistic catharsis, "symbolism will prioritize: ordinary words but will use them connotatively or with a different meaning"; the coining of new words; rare words; "unsoiled lexicon, especially in Baudelaire, Rimbaud, and, at a larger extent, Mallarmé. In the text titled "Le Tombeau d'Edgar Poe", Mallarmé eminently endorses this cathartic program by declaring:" Il faut "donner un sens plus pur aux mots de la tribu" (it is necessary to provide a **purer meaning** to the words of the tribe") Most of his poems illustrate this linguistic therapy. Such is the case of the "Ptyx" among so many others. The title itself ("Ptyx") is a semantic challenge. This poem is full of rare words: styx, nixe, onyx, septuor, ptyx**.** The following is "Ptyx", an impressive poem in terms of meaning, vocabulary, syntax, prosodic features and, more importantly, music. It perfectly subscribes to the esthetic canons of symbolism. We have, on purpose, highlighted its special vocabulary.

Ptyx

Ses purs ongles très haut dédiant leur **onyx**,
L'angoisse, ce minuit, soutient, **lampadophore**,
Maint rêve vespéral brûlé par le Phénix
Que ne se réveille pas de cinéraire amphore

Sur les crédences, au salon vide: nul **ptyx**,
Aboli bibelot d'inanité sonore

(Car le maître est allé puiser des pleurs au **Styx**
Avec ce seul objet dont le Néant s'honore.
Mais proche de la croisée au nord vacante,
un or Agonise selon peut-être le décor
Des licornes ruant du feu contre une **nixe**,

Elle, défunte nue en le miroir, encor
Que, dans l'oubli fermé par le cadre, se fixe
De scintillations sitôt le **septuor**.

In symbolism, in general, words are meticulously selected because they are the medium through which the poet aspires to suggest ethereal or celestial realities. They cease to be "algebraic signs" as in the Pythagorean vision of the universe to become "celestial signs", and, accordingly, the semes, symbols, or elements of correspondences that suggest the ethereal.

iii—A suggestive lyricism

Lyricism is a mode of expression and predominantly poetic featuring the use of exaltation of personal feelings, and passion by an author. Lyricism manifests itself through a number of criteria:

- The use of a psychological lexicon (amour, aimer, passion, colere, coeur, sentiment, joie, pleurs, bonheur, flamme, douceur, plaisir. love, to like, passion, ire,);
- French pre-positioned adjectives, that is: adjectives positioned before the noun rather than after. In French, pre-positioned adjectives (as opposed to post-positioned adjectives) have a subjective and, therefore, lyrical connotation;
- Lyrical pronouns: **Je (I), moi (me), mon, ma, mes, m' (my);**
- Superlatives **(Suprême, très grand, le plus grand, etc . . .);**
- And everything that is able to express emphasis or excess.

Lyricism is intended to communicate the effusion stemming from the psychological landscape of a particular writer who subconsciously or consciously seeks the connivance, collusion of a sensitive soul in whom he/she can confide. Originally, it was a mode of expression in which the poet was using his lyre, but since such a mode was inherently bearing an effusion of feelings, when, over time, the lyre was forsaken, lyric poetry became known as a kind of poetry bearing that effusion as well. On the basis of this explanation, it appears that that lyricism is suggestive in itself. However, when it overuses the exaltation of feelings to communicate a special effect it becomes particularly sharp and suggestive. Nevertheless, here, excess has to be found in **semantic density**, not in **lexical quantity** because symbolist poets and Mallarmé in particular are reluctant to describe, narrate, and develop profusely (against romantic lyricism, realism, naturalism, and even positivism). They prefer to synthesize because the role of the symbol is precisely to synthesize by means of the association between the concrete and the abstract. For that purpose, they use little **to mean more (Hence the propensity of creating *understatements, *metaphors, *metonymies, *asyndeta, etc . . .).** So, they utilize a pithy, concise style. Instead of naming an object, they attempt to create within us the impression or the desire of its presence or the emptiness of its absence. In *Dix-Neuvième Siècle*, Michard expounds these considerations as follows: "Mallarmé ne veut pas d'une poésie descriptive ni d'une poésie d'idées; il traduit les concepts en symboles; au lieu de nommer un objet, il tente de faire naître en nous l'impression et comme le désir de sa présence, ou le vide de son absence. The following are a few examples of suggestive lyricism.

> **Je buvais ses suaves paroles (I was drinking her sweet words); Un suprême sourire où brillât la douceur (A supreme smile where sweetness was shining).**

iv—The collage

The collage is an invention of symbolist poets but that was later on borrowed by surrealists. However, even if it is defined in *Le Manifeste du Surréalisme* (Deuxième Manifeste) by André Breton, it fundamentally remains a symbolist invention. That is why it should be named "pre-surrealist collage" or "symbolist collage" instead of "surrealist collage", which is un-factual. The collage is a technique by means of which the poet suddenly brings together two distant realities to create a powerful image. This distance can be assessed semantically, spatially or/and chronologically. In certain cases, it can be assessed on the basis of these three parameters simultaneously. The power deployed by the collage is proportional to the distance existing between these realities. It follows that the farther the realities are, the more powerful the image becomes. In the process, the image so created induces a poetic shock and, accordingly, generates poetic effectiveness. The collage, among a number of other techniques, has presided over the creation of synesthesia that we will study in chapter 3.

Examples of collage:

Un silence incolore (A colorless silence). In this instance the poet suddenly combines two distant realities: **"hearing" (silence) and "sight" (colorless);**

Le revolver aux cheveux d'or (The golden haired gun)

The word "gun" that is semantically distant from "golden haired" is brutally joined with it to create a powerful image.

v—The insertion of music into poetry, the recourse to hidden/ emotional resources of language

Music has fascinated learned minds and thinkers of all fields from philosophers to poets. Pythagoras and his disciples, the Indian poet Tagore, and mystics mentioned the music of spheres and celestial bodies. Leibniz talked about pre-established harmony, symbolist poets about harmony of correspondences/senses or treasure of correspondences (Mallarmé and Verlaine in particular). Einstein also valued music and in 1929 he confessed: "if I were not a physicist I would probably be a musician. I often think in music. I live my daydreams in music. I see my life in terms of music . . . I get most joy in my life out of my violin." (153, *Einstein a Hundred Years of relativity*). For him too, music focused on the concept of harmony. Even nowadays eminent physicists who conceived Superstring Theory talk about the "song of nature in the gentle wanderings of celestial bodies and the riotous fulminations of subatomic particles." (135, *The Elegant Universe*) to stress the harmony stemming from music. So, all these outstanding minds of the past and present concur to acknowledge that the word 'harmony' refers to one notion: music. Research has proven that it can virtually be found everywhere in the universe, used and honed to spice our lives, enrich human endeavors and, by the same token, provide them with an extra dimension. That is precisely the goal of music in symbolism. As an indispensable tool of symbolist poetry, it is employed because of its suggestiveness and symbolists believe that, under certain conditions, it is able to make us reach the transcendental (extra dimension). Its importance is mentioned in *Dictionary of Literary Terms and Literary Theory* to emphasize not only its primacy in symbolism but also and, more specifically, its power of suggestion likely to fit in specific forms of verses. *The Dictionary of Literary Terms and Literary Theory* states:

> "The music of the words provided the requisite element of suggestiveness. Verlaine in his poem *Art poétique* (1874) for instance, says that verse must possess the musical quality "De la musique avant toute chose". Such a point of view was also expressed, in other words, by Mallarmé, Valery and Rimbaud. Theory and practice led the French symbolist poets to believe that the evocativeness and suggestiveness could be best expressed by verse forms that were not too rigid. Hence the adoption of vers liberés and vers libres. Rimbaud and Mallarmé were the main experimenters of these forms; Rimbaud was the chief practitioner of the "prose poem. Such verse enabled the poet to achieve what Valery described as 'cette hésitation prolongée entre le son et le sens.' (887)

Music became one of the top priorities of symbolist poetry. Verlaine emphasized it in his *Art Poétique*. Moreover, the notion of music, free, loosen, uneven verses was intimately linked with transcendentalism and the aspiration "to attain the vision of the essential Idea". With respect to music quality, Valery considered that a verse needed to be endowed with an exceptional density honing flexibility, fluidity, music in order to obtain suggestiveness and evocativeness and make the poet access the realms of the transcendental. This standpoint was equally shared by Rimbaud and Mallarmé. The *Dictionary of Literary Terms and Literary Theory* gives further clarifications of this standpoint:

> *The attainment, in transcendental symbolism, of the vision of the essential Idea was to be achieved by a kind of deliberate obfuscation or blurring of reality so that the ideal becomes*

clearer. This, according to symbolist theory could be best conveyed by a fusion of the images and by the musical quality of the verse; by in short, a form of so-called pure poetry. The music of the words provided the requisite element of suggestiveness. Verlaine (887)

Music, as a powerful tool, was used in multifarious ways and especially under form of:

- Paronomasia, that is a technique in which one puts together words having morphological similarities by means of one or more of these prosodic or stylistic devices (imitative harmony, homometrical structure, homophonic phrases, homonymic collision (clore, clouer) see footnotes),
- Alliterations,
- Assonances,
- Rhythm,
- Metrical structure,
- Rhyme scheme,
- Silence (musical silence or silence bearing music),
- palindrome,
- Any element likely to provide repetition or a recurrent pattern because music in its raw/elemental form is nothing but repetition,
- Eutaxy, that is the meticulous arrangement of each of these elements in agreement with the perfect organization of the whole set (from Greek 'taxis' = arrangement and 'eu' = beautiful, good, well),
- All of the above and condensed in one concept: phonic mimologism.

Notes:

Imitative harmony: device through which a poet uses recurrent patterns to imitate specific sounds or spellings

Homometrical structure: use of the same metrical structure throughout a text (binary, ternary rhythm, etc . . .)

Homophonic phrases: use of words or constructions having the same sounds but not necessarily the same meaning or spelling (**mettre/metre [metr]** in French **right/wright [rait]** in English)

Homonymic or paronymic collision: use of words having a graphological kinship by brutally putting one next to another in a verse or sentence (clore/clouer) Palindrome: poem, verse or just a sentence in which one uses words that can be read from left to right indistinctively (e.g. **Laval, rotor**) to produce a specific effect, usually a musical or ludic one.

Phonic mimologism: particularly stresses phonic similarities linked with paronymic derivation, a technique formalized by Aristotle during the Greek Antiquity to create words on the basis of paronyms: words morphologically akin in terms of their etymology. Phonic mimologism systematically assembles words by taking into account their musical kinship. Through this process, the poet elicits amazing effects from the suggestive power of sounds. This is particularly remarkable in Mallarmé who is regarded by the most authoritative critics of symbolism (Claude Pichois, Llyod Austin, Jean Pommier, Jean Peyre) as one of the greatest symbolist poets. Thus, we choose to use a few instances from his poems. For instance, his sonnet "Ptyx", whose rhyme schemes is **xy, ix, ix**, turns out to be outstanding with respect to the creation of an atmosphere combining magic, supernatural, a sense of eeriness and, above all, music. By means of the music so created the poem becomes infused with the symbolism of the mystery enhanced by the fact that "x, y" stand for unknown concepts in mathematics. I strongly believe that the rhyme scheme (yx) suggests this symbolism of the mystery, that is: the unknown, which implicitly puts the reader in touch with it. Here, we will just mention paranomasia, silence and eutaxy since they are some of the most used musical elements in symbolism and Mallarmé's poetry in particular.

Example: Ptyx

Ses purs ongles très haut dédiant leur **onyx**,
L'angoisse, ce minuit, soutient, **lampadophore**,
Maint rêve vespéral brûlé par le Phénix
Que ne se réveille pas de cinéraire amphore

Sur les crédences, au salon vide: nul **ptyx**,
Aboli bibelot d'inanité sonore
(Car le maître est allé puiser des pleurs au **Styx**
Avec ce seul objet dont le Néant s'honore.

Paronomasia

In *Ptyx* paranomasia is, on the one hand, expressed by the collusion 'sonore' and 's'honore' (perfect homonyms) in the fourth quatrain, which is factually an extension of the second. On the other hand, it is formulated by the rimes in yx/ix with onyx, phenix and especially with the *minimal pairs: Styx/ ptyx. 'Minimal pairs', a linguistic notion also called 'quasi-homonym pairs' is a group of words whose spelling is identical except for one *phoneme. Here it is the case of | s | and | p | respectively in S̲tyx and p̲tyx. There are also 'lampadophore' and 'amphore' that constitute a paronomastic set. If we take into account the meticulous choice of words, their phonological or phonetic collusions, the prosodic and *euphonic effects, we are prone to realize the extent of Mallarmé's command of music and prosody. These few instances about the occurrences of paronomasia show that Mallarmé was steeped in music and prosody.

Likewise, here "Le Sonnet du Cygne," another of his poems, carries this musical quality a step further through paranomasia. It includes one, some or all of the following technical devices: imitative harmony, euphonic effects generated by echoes produced by rhyme schemes, homometrical structure, homophonic phrases, homonymic collision (e.g. clore/ clouer, "assigne/le cygne" in the poem). Here, the paranomasia is generated by the rhyme scheme (ui), (ivre); imitative harmony (goni) vs. (leni), (ivre) vs. (givre), (assigne) vs. (lesigne)

Le Vierge, le vivace, le bel aujourd'**hui**
Va-t-il nous déchirer avec un coup d'aile **ivre**
Ce lac dur oublié que hante sous le **givre**
Le transparent glacier des vols qui n'ont pas **fui**

Un cygne d'autrefois se souvient que c'est lui
Magnifique mais qui sans espoir se délivre

Pour n'avoir pas chanté la région où vivre
Quand du stérile hiver a resplendi l'ennui

Tout son col secouera cette branche agonie
Par l'espace infligée à l'oiseau qui le nie,
Mais non l'horreur du sol où le plumage est pris

Fantôme qu'à ce lieu son pur éclat assigne
Il s'immobilise au songe froid du mépris
Que vet parmi l'exil inutile le Cygne.

The rhythm of verses and the harmony of sounds (*paronomasia, homophones, homometry, etc . . .) obtained by the eminently organized rhyme scheme causes this sonnet to stand out sub specie aeternitatis: (h**ui**, f**ui**) (**ivre**, g**ivre**) (l**ui**, enn**ui**) (del**ivre**, **ivre**) (ago**nie**, **nie**) (**pris**, **mépris**) (as**signe**, **Cygne**). It virtually gives way to Cratylism, that is the warding off of the arbitrariness of language. Music thus causes the language of the poem to be motivated. It endows it with a semantic quality, symbolism. It cryptically helps the analyst—through its rhythm, artful organization of sounds, and the rhyme scheme—to read the soul, the subconscious, the dreams, and aspirations of the poet. The sonnet is finally a lively psychological landscape.

Silence/pause bearing or infused with music

    Sainte
A ce vitrage d'osten**soir**
Que frôle une harpe par **l'Ange**
Formée avec son vol du **soir**
Pour la délicate pha**lange**

Du doigt que, sans le vieux **santal**
Ni le vieux livre, elle ba**lance**
Sur le plumage instru**mental**,
Musicienne du si**lence**.

Here, the poet exploits the hidden resources of words, pause, and absence of words, echo effects to create music. He becomes a thaumaturge, a miracle monger in the sense that he can virtually use anything to create music. He is factually able to create music out of silence/pause. The pause and the subsequent silence after each rhyme create an echo whose effect, generated by imitative harmony, is responsible for the creation of music. This is enhanced by the use of a stylistic device named 'asyndeton' especially in the first stanza. The asyndeton refers to the ablation or suppression of connectors or very little use of particles of connection (et, mais, or, ou, car, ni, quand, comme, etc.). De facto, in stanza 1 each verse functions as if it were a prosodic entity regulated by an homometrical structure, a melodic unit stressed at the end of the rhyme. Each rhyme and its phonological structure and pause generate echo and, by the same token, music. Additionally, each verse is sublimely governed by a meticulous choice of words on the basis of sonorities. That is why the pause between verses and rhymes contribute to produce music. The poem *" Sainte "* bears the seal of the poet as a thaumaturge in the last verse. Indeed, he/she proves to be the "musicien/ne) du silence": the musician who creates music out of silence. The poem is therefore charged with a metalinguistic quality since it poeticizes, defines both itself and the poet.

Eutaxy

Mallarmé's poetry is a paradigm of prosodic and musical mastery. That is why most of his texts were set to music. Besides, his poetry reflects, in general, a very highly refined organization because everything is so artfully conceived that it became a pretext to infuse texts with music: balance and symmetry of stanzas; parallelisms; their homometrical structure; words and sounds minutely chosen to generate musical effects on the basis of their hidden prosodic and linguistic resources (**assigne/cygne, hui/fui, givre/ivre, pris/mepris**) in *"Le Cygne"* (**ostensoir/soir, ange/phalange** [ãʒ]**, balance/silence** [balɑ̃s] [silɑ̃s]) and *"Sainte";* silence/pause generating echoes at the end of each verse and, by the same token, producing music; high frequency of vowels rather consonants (vowels, unlike consonants, are melodious sounds. Thus, they are much more endowed with musical qualities), a very subtle endeavor to attune the signifier with the signified, a deliberate effort to deny any stylistic randomness in favor of semiotic effectiveness in such a way that the overall organization of the poem is systematically geared towards a monolithic unity stylistically and semantically. Such a pattern is especially achieved in *Le Cygne* and *Sainte*. "Sainte" in particular corroborates a linguistic rigor worthy of a masterpiece. The last verse functions like a self signature because, through it, the poem seems to assert itself as an epitome of musical achievement generating music out of silence: "la musicienne du silence". Consequently, Mallarmé's poetry is a model of musical excellence and eutaxy. This ambition of excellence climaxed into his creation of a new poetic syntax.

vi—The creation of a new poetic syntax

The symbolist poet, and Mallarmé in particular, will recur to the hidden emotional resources of language by means of music and the atomization of the verse. By "atomization of the verse" we mean its breaking down into small linguistic units through a process of restructuring, disarticulation or dislocation. Mallarmé will dislocate the sentence/the verse by disjoining the verb from the subject, the infinitive from the auxiliary, by proliferating appositions—often located before the word to which they refer—, ellipses, inversions, and periphrases. There will also be in Mallarmé the propensity of recurring to certain stylistic devices such as: ***asyndeton, *hyperbaton, and *understatement** in symbiosis with this phrastic dislocation. **The asyndeton** is a stylistic device characterized by the suppression of conjunctions and any particle likely to suggest the connection of words, ideas, combination or junction in a sentence or a verse (**and, or, but, for, either . . . or, neither . . . nor). The hyperbaton** refers to the inversion of the structure of the sentence/verse to emphasize an idea, a notion or a concept (**"Blue, the sky is"** rather than **"The sky is blue"; "That, I will do"** instead of **"I will do that"**). With respect to **the understatement**, it is a technique conceived to mean more than one factually says (**"She is not ugly"** for **"she is extremely beautiful"**). Most of the time in an understatement one uses a few words to develop more meaning. Therefore, it has a semantic density, and quality.

All this heuristic work elaborated by the atomization of the verse, the propensity to recur to certain stylistic devices definitely lead to the creation of a new poetic syntax which, somehow, reminds of the word order in classical Latin. In *Dix-Neuvième Siècle*, Michard glosses on this artfully elaborate syntactic process. He acknowledges that Mallarmé's outstanding research focused on the structure of the sentence: #"C'est sur la structure de la phrase que va porter son effort. Il la disloque en écartant le verbe du sujet, l'infinitif de l'auxiliaire, en multipliant les appositions (souvent placées avant le mot auquel elles se rapportent), les ellipses ou au contraire les périphrases. Cette désarticulation n'est nullement livrée au hasard: elle figurera par exemple, dans un naufrage, le désordre des débris épars sur les flots (A la nue . . .) "(*Dix-Neuvième Siècle*, 531)

The power of suggestion culminates into the use of vers libres (free verses), vers impairs (uneven verses), and vers liberés (loosen verses). All these three types of verses are designed to give the poet a greater opportunity and flexibility to develop ideas, use music, cogently express his inner feelings and, accordingly, suggest more. In this respect, the rhythm is no longer conditioned by syntactic principles or conventional principles of prosody (**rhythm as a mere prosodic feature => external rhythm**). It is henceforth conditioned by the psychological drives of the poet's ego (**rhythm as a psychological feature** or a prosodic feature but contingent upon the very poet's psyche => **internal rhythm**). Therefore, the adoption of the vers libres, les vers impairs and vers liberes by symbolists provide them with a special opportunity to exercise their flexibility and enhance their suggestiveness and creative power at a higher level. All these three types of verses prove to be a deliberate reaction against the despotism of French classical versification whose very strict rules finally thwarted or at least reduced creativity. The liberation of the verse started with Romantic poets and mainly Victor Hugo, the leader of Romanticism. In *La Preface de Cromwell*, regarded as the manifesto of Romanticism, Hugo rebelled against the rules of French classical versification and decided to change the structure of the alexandrin verse. He stated: "J'ai disloqué ce grand niais d'alexandrin et mis un bonnet rouge sur le vieux dictionnnaire" ("I have dislocated this great silly alexandrine and put a red cap on the old dictionary"). Here, "dislocating the alexandrine" refers to the modification of its metrical structure. Classically, it used to be binary (a rhythm of two metrical units or two sequences divided by a pause called "caesura") and composed of two parts called hemistiches. Each hemistich had six syllables. Therefore, the alexandrine also called "dodecasyllabic verse (i.e. consisting of twelve syllables—:> from the Greek prefix "dodeca" meaning twelve) was composed of twelve syllables. The following are 1—the classical alexandrine verse, 2—the romantic alexandrine, 3—the symbolist free verse, which enables to see the differences between them.

1—The structure of the classical alexandrine

Example: "Et/Ro/me/ t'a/ choi/si/, I I je/ ne/ te/ con/nais/ plus" (Corneille)

6 syllables	Caesura	6 syllables
1st hemistich		2nd hemistich
1st rhythmical unit		2nd rhythmical unit
		=>Binary Rhythm: 2 rhythmical units

Then, Hugo and his followers decided that the alexandrin would no longer have a binary rhythm (external, conventional). Its rhythm will be internal, that is, conditioned by the sole mood, the psychological disposition of the poet (ternary (3 pauses) or even quaternary (4 pauses) according to the poet's own free will and mood). With respect to "putting a red cap on the old dictionary", it alluded to the fact that poets were no longer obliged to make use of a classical lexicon for the sake of decency (rules of "bienseances") and lexical decorum. They now assigned themselves the freedom of using an extended vocabulary even that of the revolted people. "Bonnet rouge" symbolized the revolted people. Thus, the liberation of the French verse was a gradual process. It started with French Romanticism fuelled by Hugo and his peers and reached its paroxysmal stage with French symbolism through the vers libéré (the freed verse). Baudelaire, Rimbaud, Verlaine and Mallarmé in particular catalyzed this process of liberation.

2—The structure of the Romantic alexandrine

Example: "O temps/, suspends ton vol/! /ll Et vous/, heures propices/," **(Lamartine)**

I	II	III	IV
1st hemistich	Caesura	2nd hemistich	
2 rhythmical units		2 rhythmical units => Quaternary Rhythm: 4 rhythmical units	

3—The structure of the symbolist vers

Example: "Souvent/, pour s'amuser/, II les hommes d'equipage/" **(Baudelaire)**

1st hemistich		Caesura	2nd hemistich
I	II		III
2 rhythmical units			**1 rhythmical unit => Ternary Rhythm: 3 rhythmical units**

- The vers libres are free verses in the sense that different syllabic lengths and also called heterometric verses (i.e.: of different measure/length from Greek "hetero": "other, different" and "metro": measure) were used within the same stanza.
- The vers impairs are verses having an uneven number of syllables like: l'hendecasyllable (11 syllables); the enneasyllable (9 syllables), etc . . .
- The vers libéré is a verse completely freed from any metrical or rhythmic restriction (caesura, counting of syllables and mute e, etc . . .). However, in some poets, rhyme and some sense of syntactic unit may persist. The only notion that is taken into account is the internal rhythm, that is the psychological rhythm of the poet. In light of all these considerations, it appears that the liberation of the verse has significantly enabled symbolist poets to excel in terms of suggestiveness, creativeness and, above all, quality. *The Dictionary of Literary Terms and Literary Theory* endorses this particular aspect:

"Theory and practice led the French symbolist poets to believe that the evocativeness and suggestiveness could best be obtained by verse forms that were not too rigid. Hence their propensity of utilizing the vers liberes and vers libres (q.q.v). Rimbaud and Mallarmé were the main experimenters in these forms; Rimbaud the chief practitioner of the 'prose poem'(q.v). Such verse enabled the poet to achieve what Valery described as 'hésitation prolongée entre le son et le sens

In addition to the power of suggestion, the awareness of the style is part of the esthetic canons of symbolism.

III—"THE AWARENESS OF THE STYLE"

In symbolism "the awareness of the style" is so important that it deserves to be examined. It is a concept conceived as the ability for the poet to create by the sheer power of his will, that is, without the assistance of inspiration. In other words, by virtue of this aptitude, there is a far greater primacy of composition over inspiration. Accordingly, the poet and his material are no longer subordinated to inspiration but to the density of his own composition. It appears that most genuine, bona fide symbolist poets are driven by this fundamental preoccupation: to create the beautiful by the sole power of their will as they are obsessed by their ideal of absolute poetry. Thus, they will hone the heuristic instrument enabling them to achieve this ideal at the highest level. De facto, the preoccupation of the symbolists who epitomized the awareness of style was to invent a special language that is, a language that can compensate the imperfection of ordinary language. Such a preoccupation had an underlying poetic function. Barthes comments on it in an article titled "Proust et les noms":

> (La fonction poétique) au sens large du terme se définirait ainsi par une conscience cratyléenne des signes, et l'écrivain est le récitant de ce grand mythe séculaire qui veut que le langage imite les idées et que, contrairement aux précisions de la science linguistique, les signes soient motivés. (7)

In *Le Symbolisme*, Svend Johansen refers to the awareness of the style as a capital asset. For him, it is the real benchmark by which one can distinguish real symbolists and those who are not. His reflections are illuminating:

> *La conscience du style chez les symbolistes fait de leurs oeuvres un excellent objet de recherche, si l'on désire surprendre la création poétique dans sa fonction. Leur degré d'importance à cet égard dépendra du point auquel le problème du style [...] a préoccupé le poète. Une classification des symbolistes sur la base de ce degré d'importance peut permettre d'écarter plus facilement les poètes qui, seulement pour des raisons de contemporanéité ont reçu le nom de symbolistes. Les poètes qui seront écartés sont ceux dont la conscience du style n'a pas été assez forte pour prendre la forme d'un complexe* (20).

The following is the translation of these important reflections:

> *The awareness of the style in symbolists causes their works to become an excellent heuristic object when one wishes to find out poetic creation in its function. Their level of importance will be contingent upon the extent to which the issue of the style has preoccupied the poet [....] A classification of symbolists on the basis of this level of importance may enable to discard the poets, who, on the sole grounds of contemporariness were labeled "symbolists". The poets who will be discarded are those in whom the awareness of the style has not been cogent enough to be translated into a kind of complex (20).*

In light of these criteria, it is possible to distinguish transient symbolists and real symbolists. Jean Peyre in *Qu'est-ce que le Symbolisme?* considers that Mallarmé, Valery, Rimbaud, Baudelaire belong to the second group, that of real symbolists. Other authoritative critics and especially Claude Pichois, Jean Pommier, Llyod Austin also consider that they are among the best representatives of symbolism.

IV—THE USE OF SYNESTHESIA

The last major esthetic canon symbolists used is synesthesia. Synesthesia is a holistic phenomenon shared by poetry, neuroscience, psycholinguistics, semiotics, to mention but a few. It stems from the conjunction of two or more different sensory perceptions triggered by specific stimuli. It is generated in the left hemisphere of the human brain and poetically translated into language (this aspect will be developed later). Synesthesia is governed by the psychological law of totality, according to which all sensory modalities can be transcended and synthesized to constitute an organic unity. In its highly refined stage, synesthesia can lead to Cratylism, that is: the aspiration to reduce or ward off the arbitrariness of language by utilizing sophisticated techniques conceived to motivate the relationships between *the signifier and signified. Given that synesthesia is complex and extremely important within the framework of symbolism, I will devote a whole chapter to it and provide it with a more complete definition. Synesthesia will be studied with a semiotic approach.

SUMMARY OF CHAPTER II

Symbolism goes from 1850 to the end of the century. It stems from platonic idealism and reacts against romanticism, the Parnassian school, naturalism and positivism. All these schools try to **describe** reality without success because it is complex, surreptitious. It constantly changes. Thus, the proper way to deal with it is to **suggest** how it appears through specific canons. Such is the approach adopted by symbolism. Additionally, symbolism aspires to materialize absolute poetry heralded by Baudelaire's *Fleurs du Mal* published in 1857. Its theoretical program systematically disqualifies whatever is alien to poetry. It entails a number of ambitious technical possibilities designed to reach this poetic ideal: the transcendence of reality to show hidden connections between the material and the immaterial universe; the power of suggestion governed by specific technical devices: use of symbols; the quasi—systematization of connotation to avoid linguistic prostitution; the adoption of a special, erudite vocabulary; the collage; music, uneven verses; free verses; the "awareness" of the style or the capacity for the poet to create by the sole power of his will (composition) rather than that of inspiration; and, most importantly, the use of synesthesia leading to the attempt to ward off/reduce the arbitrariness of language by bridging a gap between the signifier and signified. Most French symbolists are poets and sophisticated scholars of poetry (Baudelaire, Rimbaud, Mallarmé, Valery) who seek the hidden emotional resources of language to make it eminently powerful, dig its lexical and semantic abyss, which culminates into the creation a new form of language: synesthesia.

CHAPTER III

SYNESTHESIA AS THE ACME OF SYMBOLISM THROUGH CRATYLISM, ITS TOOL OF HERMENEUTIC INVESTIGATION: SEMIOTICS

> "**NIHIL EST IN INTELLECTU QUOD NON PRIUS FUERIT IN SENSU**" (Thomas Aquinas borrowed this aphorism from Aristotle. It became the motto of *empiricist philosophers Hume, Locke and Hobbes)
>
> "**NOTHING IS IN THE INTELLECT/INTELLIGENCE THAT WAS NOT FIRST IN THE SENSES—ALL OUR KNOWLEDGE IS MEDIATED THROUGH THE SENSES**"

Synesthesia is one of the major components of symbolism. As a matter of fact, it is one of the eminent esthetic canons of symbolism. Each pure symbolist poet arrogates himself the right of employing it or at least dallying with it in his work. Henry Peyre, a remarkable scholar of symbolism, acknowledges the capital importance of synesthesia. In his book titled *Le Symbolisme* he states: "L'élément salué" comme symboliste avant la lettre fut donc l'usage des synesthésies" ("the element appraised as symbolist before the official formation of symbolism was definitely the usage of synesthesia"). Mallarmé buttressed this viewpoint in *Symphonie Littéraire*. *He indeed* featured the preponderant position of synesthesia in this text. He declared: *"Le trésor profond des correspondances, l'accord intime des couleurs, le souvenir du verbe antérieur, la science mystérieuse du verbe sont les qualités du symbolisme."* (**27**) By "trésor profond des correspondances" Mallarmé implicitly referred to synesthesia, also called "horizontal correspondences" or correspondences between different sensory perceptions as opposed to vertical correspondences, that is: correspondences between the material and the ideal or supra-terrestrial universe. These two considerations definitively materialize a consensus around the capital importance of synesthesia within symbolism. Accordingly, studying symbolism pre-requires the necessity of studying synesthesia as well. However, for the sake of conducting a meticulous and coherent analysis on synesthesia it is useful to delineate its historical background.

1—PITHY HISTORICAL BACKGROUND

One of the first official records of synesthesia goes as far back as the Greek Antiquity and can be found in Plato's dialogue named *Cratylus*. In it, there are two characters: Cratylus and Hermogenes involved in a discussion. Cratylus argues that sounds have a meaning and, for this reason, are referential whereas Hermogenes disagrees with him. Through this discussion Cratylus tacitly refers to the arbitrariness of language and the implicit possibilities of warding off such arbitrariness by motivating the signifier(s) by means of synesthesia (Here, signifiers are sounds endowed with meaning(s)). Cratylus, therefore, indirectly laid the foundations of phonosemantics, a branch of linguistics designed to ponder on the possibility for sounds to express meaning(s). It follows that *Cratylus* might be regarded as one of the first official adumbrations of phonosemantics and synesthesia. Another reference of synesthesia is provided by Pythagoras. He discussed about it in the 6th century B.C. and it seems that, later, in the 4th century B.C., Aristotle did likewise. Much later, in the seventeenth century Locke wrote *Essay Concerning Human Understanding*. In his essay he told about what happened to a blind man. He received a shock in his head while he was in the company of his friend. Then, when he described the effect of the shock to his friend he said that it was scarlet like the sound of a trumpet. This was a pure manifestation of a synesthetic phenomenon: association of different sensory modalities (visual modality -> **scarlet** + auditory modality -> **sound of a trumpet**) to constitute an organic unity by means of the psychological law of totality. Leibniz reported another case of a blind man who said that he really understood what scarlet was. According to him, the scarlet color was like the sound of trumpet. Nevertheless, it was factually in the second half of nineteenth century that interest in synesthesia reached its peak due to the influence of artists, writers, philosophers, and scientists. A conjunction of two basic influences sparked the belief in the factualness of synesthesia: writers/artists, and the philosopher Swedenborg who lived in Sweden one century before but whose influence was still very strong. Indeed, numerous writers and artists intuitively realize that there is a keyboard of senses within which certain sensory perceptions are transposable to the extent that they can tangle, intertwine, merge, or at least correspond with each other. In *You Forever* Dr. Lobsang Rampa attested to the existence of this symbolic Keyboard. We will examine it later on. Artists, writers, philosophers and scientists also believed in the correspondences between the arts and aspired to invent the perfect art, that is, an art able to synthesize all the possibilities of human creativity: music, songs, choreography, visual representations, touch, mimicry, proxemics (art and language of distance), kinetics, etc . . . this art was coined "opera". Some writers endowed with a mystical and spiritualist temperament found in it a source of inspiration to conceive and concretize a form of artistic communication likely to transpose the principles of the opera into literature, painting or other arts in order to touch the soul of the beholder. In *Synesthesia, Classic and Contemporary Readings*, Harrison and Baron-Cohen provide an explanation of this artistic, spiritualist and mystical background. They show the influence of Wagner's opera on composers and painters like Scriabin and Kadinsky, poets like Baudelaire and Rimbaud. Such influence had inspired them the elaboration of synesthesia. Their reflections are pertinently expressed in this statement:

> The composer Alexander Scriabin and painter Wassily Kandinsky were almost certainly aware of the efforts of Rimbaud and Baudelaire to link the senses and seem to have been aware of another's work; both were also influenced by

> *the operatic work of Richard Wagner (1813-83). Wagner saw opera as the ideal medium for conveying his thoughts and ideas and hoped 'to touch the Christianity within us' . . . Kandinsky, in his written work* <u>On the Spirituality in Art</u>, *takes up Wagner's suggestion that it is possible to touch the inner spirituality through the arts. Kandinsky appears to have possessed a certain envy of music, whilst being almost totally abstract, can successfully conjure visual I mages (cf: Debussy's Clair de Lune, and Wagner's overture to Das Rheingold) [. . .] Kandinsky's move towards total abstraction in his work seems to have followed [. . .] from his desire to imbue his work with a synaesthesic quality. His intention was for canvases. This evocation of an auditory dimension to visual representations was a move towards Kandinsky's ultimate aim of creating the gesamtkunstwerk ("total art work") (9-10)*

This total art work was called "opera". In addition to the aspiration to create the perfect art, there was a significant interest in the theory of correspondences developed by the Swedish philosopher Swedenborg. Indeed, most thinkers, artists and writers were influenced by Swedenborg and his theory of correspondences. In his book *Heavenly Secrets* this philosopher talked about invisible connections linking all the elements of creation. By them he implicitly meant what will be termed 'synesthesia' later. Thinkers, artists, and scientists strongly believed that just as there are correspondences in the universe, there are correspondences between the arts, and the five senses: sound, sight, smell, touch, and taste. Kandinsky, Scriabin, Nabokov, Baudelaire, Rimbaud, Mallarmé, Huysmanns, Rodenbach and most symbolist poets who were probably *synesthetes themselves subscribed to the theory of correspondences and the unity of senses. It appeared that this phenomenon first accredited by artists, writers, and philosophers finally got the attention of scientists such as: Leibniz and Newton, among others, in the eighteenth century. Unfortunately, in the nineteenth century behaviorism became the dominant epistemological paradigm and, later, in the years 1920' 1930's, behaviorism was greatly influenced by logical positivism, which scientifically conditioned the future of synesthesia.

Influence of logical positivism

Logical positivism was a a system of thought formalized by a team of scientists and philosophers (Kurt Godel, Gustav Bergmann, Rudolf Carnap, Hans Hahn, Olga Hahn-Neurath, Otto Neurah Theodor Radakovic were just a few of them) called the Vienna Circle. Logical positivism synthesized *empiricism, *rationalism and *epistemology. Its key idea was the systematization of the principle of verifiability in light of methods used in math, logic and empirical sciences and apply it to every field of human knowledge. Almost any intellectual quest aspiring to be scientific had to comply with the principles of behaviorism and logical positivism. These principles were: observation, analysis and, most importantly, verifiability or verificationism. On page 237, the *Dictionary of Important Theories, Concepts, Beliefs, and Thinkers* provides further clarification on this specific point:

> The logical positivists maintained that all intellectual inquiry should be held to the same standards as scientific investigation. Paramount in this conception was the principle of VERIFIABILITY(sic), which states that for a proposition to be meaningful, it must be not only logically consistent but susceptible to empirical verification—not necessarily proven, but at least able to be tested. By this criterion, all METAPHYSICAL (sic), religious, and ETHICAL (sic) statements were banished as unverifiable and therefore meaningless; only what can be evaluated scientifically was significant. (237)

The result of the osmosis behaviorism-logical positivism prevented synesthesia from soaring quickly because it was considered as pertaining to the realm of psychology, which was that of mental processes. Unfortunately, according to logical positivism, any phenomenon purported to be mental or psychological was not subject to scientific scrutiny or verifiability. Given that it was not possible to see or fathom what was occurring inside the mind and the human brain, mental processes were neither observable, analyzable, nor verifiable. In the final analysis, because synesthesia involved mental processes and subjective experiences, it was disqualified as unscientific and dismissed from scholarly research. Subsequently, the interest in synesthesia gradually declined. However, against all the odds, behaviorism and logical positivism turned out to be propitious and beneficial to synesthesia and researchers. Indeed, it influenced scholars of this field by intellectually and vicariously coercing them to stick to the principle of verifiability to achieve a sound dose of scientific credibility. Consequently, researchers understood that the assertion of synesthesia as a genuine, bona fide discipline would require them to reject a psychological approach and condition their research to scientific/empirical verification. Fortunately, with the end of the nineteenth century and the beginning of twentieth century, the impressionist movement started, which rekindled the interest in synesthesia and mental states in general. Later on, during the period between the two World Wars, surrealism and neurology blossomed and both of them boosted the development of synesthesia in such a way that they helped it to become a full-fledged discipline.

Influence of surrealism and the development of neurology

The surrealist movement emphasized all phenomena akin to the unconscious, alchemy, extrasensory perceptions, and above all spiritualism and forces dormant in man. Therefore, surrealism re-triggered the interest in synesthesia. However, it is really with the development of neurology (when it ceased to back its analysis on psychology) and the physiology of the human brain that synesthesia emerged from a psychological realm to a mere scientific one. Nowadays, researchers like: Simon Baron, Cohen, Ramachandran, Richard Cytowic, to mention but a few, have done a remarkable work on synesthesia by assessing and screening what occurs inside the human brain without recurring to surgery. Three basic sets of methods and scientific devices have enabled to scientifically prove the factual existence of synesthesia: Positron Emission Tomography (P.E.T.), Functional Magnetic Resonance Imaging (F.M.R.I.), and the test/report on color grapheme synesthesia. We will provide substantial information on these methods farther. Still, it is noteworthy that synesthesia is a relatively new field and any research and finding are provisional and should be completed, amended or readjusted as new discoveries will be available over time. Moreover, asserting the factual reality of synesthesia and the ability for the human brain to generate poetry presupposes the possibility to know what occurs inside it. That is why the cogent analysis of synesthesia requires the conjunction of several fields: neuroscience and poetry/semiotics, linguistics, and psychology. For this reason, we predominantly utilize notions pertaining to these areas of knowledge to factually substantiate synesthesia. It follows that researchers should work in multidisciplinary teams or at least seek support from other specialists also interested in synesthesia (semioticians, linguists, poets, psychologists, cognitive scientists, etc . . .) to have far better results. The fact of the matter is that nobody can claim a monopoly of this "territory" and each specialist should feel proud of sharing it with others since it definitely encompasses many fields of human knowledge. The cursory historical background thus provided leads us to key—questions: what is synesthesia and how does it concretely manifest itself?

A—**DEFINITION OF SYNESTHESIA**

The word "synesthesia" stems from Greek "sunaisthesis", which means: "combination of sensations", "simultaneous sensations". The prefix "συν"("sun") means: "together, combination, conjunction" and the root αἴσθησις ("aisthesis") designates "sensations, sensory perceptions". As such, "synesthesia" is the opposite of "anesthesia": that is the absence/suppression/elimination of sensations. The prefix "a/an" designates the notion of absence, suppression, ablation, elimination, and the root "aisthesis" refers to "sensations". Thus, the word "synesthesia" designates both a faculty, a technique, and a phenomenon focused on the combination of two or more different sensory perceptions triggered by specific stimuli. These sensory modalities combine by creating an organic unity by virtue of the psychological law of totality which posits that when two or several sensory modalities are intimately linked, there occurs a phenomenon of transcendence by means of which one single inter-sensory modality emerges by transforming all the others into one monolithic unity accessible to all of them. Synesthesia deals with cross-modal sensory perceptions/cross-modal metaphors also called inter-sensory metaphors or transmodal metaphors. The denomination "cross-modal or transmodal" suggests the idea that two or more sensory perceptions/modes of perception such as sight and hearing, smell, taste and touch "cross" their anatomic boundaries and merge with others. It also suggests the perception of reality within the framework of symbolism. Within it, reality is perceived as monolithic, unified or holistic rather than fragmented or divided. Cross-modal metaphors and their *syntagmatic context are instrumental in the generation of a highly refined form of poetic expression. The construction "syntagmatic context" designates the relationships a word develops with others sequentially, that is, linearly within a sentence (in the horizontal axis). We find synesthesia when dealing with peculiarities experienced by synesthetes (that is: individuals endowed with the faculty, technique of synesthesia) such as: hearing colors; seeing sounds, colors of sounds; seeing specific colors inherently associated with specific letters, days, months, weeks, years; smelling sounds; tasting colors or sounds, to mention but a few. Synesthesia is highly suggestive and endowed with a poetic quality. For instance, in the construction "a green sound" we can assess the expression of synesthesia obtained through the combination of a **visual** sensory perception (**green**) and an **auditory** sensory perception (**sound**). Likewise in "a yellow shout" we have the manifestation of a synesthetic construction composed of a **visual** sensory perception (**yellow**) and an **auditory** sensory perception (**shout**). Here, since the visual and the auditory mode of perception coalesce, it can be inferred that a synesthete can simultaneously see and hear a sound. It appears that synesthesia is factually a complex phenomenon. Indeed, it involves poetry, semiotics, physics, and cognitive sciences. Because of this complexity, literature and semiotics by themselves cannot grasp or explain its whole density and semantic ramifications. De facto, it has a multidisciplinary status. That is why its explanation and understanding require the conjunction of cognitive sciences especially the anatomy and physiology of the human brain, neurology, psychology, linguistics, literature, semiotics, even genetics (because researchers consider that it is hereditary) and physics. For a methodic study of synesthesia, it is necessary to present its different types.

B—**TYPOLOGY OF SYNESTHESIA**

There are many types of synesthesia and it would be tedious or even impossible to provide an exhaustive list. This difficulty will be maximized by the fact that some types are not yet discovered by scientific research. However, we can mention a number of them with these combinations: color/grapheme, sight/hearing, taste/shape, taste/sound, tact/smell, hear/smell. Some scholars also reported orgasm/color (Dr. Hugo). The phrase *'united senses of the mind'* is a wonderful metaphor used by Dr. Hugo to suggest this *'unity in diversity'*. It can involve the conjunction of two (bimodal), three (trimodal), four (tetramodal—very rare) or five sensory modalities (possibly but extremely rare).

For the sake of simplification, we can distinguish four major types of synesthesia: constitutional synesthesia (also called developmental synesthesia or idiopathic synesthesia by researchers as Dr. Simon Baron and Cohen in *Synaesthesia, Contemporary Studies)*; pseudo-synesthesia, literary synesthesia, acquired synesthesia or pathological or non idiopathic synesthesia, and color-grapheme synesthesia.

- Constitutional synesthesia is acquired naturally. It is inherent in individual who were born synesthetes. Such individuals are naturally endowed with this faculty. They consistently see the same colors associated with the same letters or sounds.
- Pseudo-synesthesia designates the kind of synesthesia experienced under the influence of drugs such as marijuana, hashish, or LSD. The consumption of such drugs brings about the temporary merging of two or more different sensory modals/ psychic areas of the brain and the subject transiently becomes a synesthete. Still, this form of synesthesia does not last long, for as soon as the influence of drugs fades away the subject loses his/ her "ability". Moreover, when he/she is still under the influence of drugs, the colors, sounds or tastes that he/she perceives are inconsistent and fuzzy. They can change ceaselessly whereas those perceived by born synesthetes always remain unchanged, clear, and unequivocal.
- Literary synesthesia fall under the category of cross-modal metaphors or tropes and used by poets, writers who are usually synesthetes themselves. From a merely artistic and semiotic viewpoint, synesthesia is subtle and proves to be a very rich and highly refined form of literary expression. In *Synesthesia in Perspective*, Noam Sagiv endorses this viewpoint and believes that: "Synesthesia might represent a basic mechanism for the development of metaphors, but in a more vivid form [. . .]" **(27)**
- Acquired synesthesia is induced accidentally. If a portion of an individual's brain is damaged, some of his/her cerebral areas may interfere with another. As a result of this, the individual becomes

a synesthete. In this case, synesthesia becomes pathological. In the same vein, we can find a transient type of synesthesia with people suffering from epilepsy. This is due to the malfunction of the brain caused by a deficiency in electrical (electrical storm) or electromagnetic supply, inducing a tangling of specialized psychic areas of the brain. In such a case, it might be possible to treat them by readjusting the supply of the electricity in the brain either through specific yoga or *ki exercises. However, in each type of synesthesia, it is possible to appreciate the relationships between thought and language, the interaction between the function of the brain, thought and that of language. This is precisely the field of psycholinguistics because it deals with the relationships between thought and language.

- Color-grapheme synesthesia is a type of synesthesia in which letters, numbers, days, weeks, months, years are always associated with specific colors, shapes, textures and sometimes, mood and genders. It is one of the commonest type of synesthesia and this was pertinently demonstrated by Ramachandran & Hubbard in their article *The Emergence of the Human Mind: Some Clues from Synesthesia* by means of the theory of adjacency and synaptic plasticity (152). As a matter of fact, synapses are plastic and, accordingly, flexible, malleable. Therefore, under certain conditions (cases of emotion/ activation of the limbic system), they can tangle, move easily. Additionally, the brain area responsible for colors and the one that is responsible for letters are vicinal, next to one another. Consequently, the plasticity and proximity of their synapses provide a plausible explanation of their higher possibility to tangle and merge to cause the psychic areas to which they belong to merge as well in order to produce color-grapheme synesthesia **(see the brain areas concerned in the following photograph).**

Cross-activation

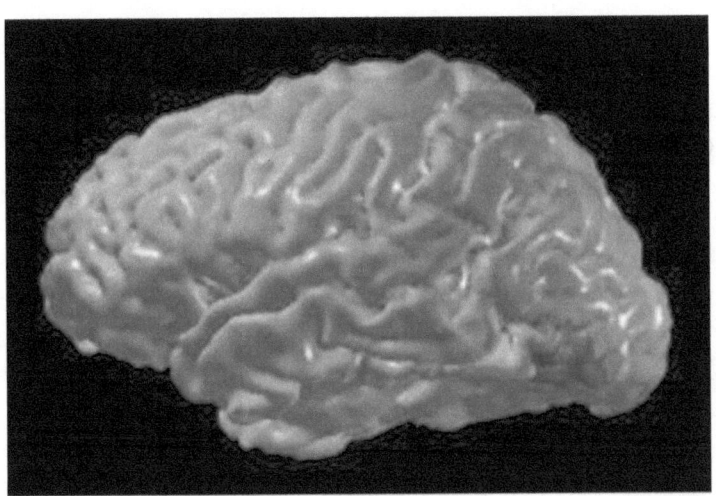

Regions thought to be cross-activated in grapheme-color synesthesia (from Ramachandran & Hubbard 2001).

Since regions involved in the identification of letters and numbers lie adjacent to a region involved in color processing (V4), the additional experience of seeing colors when looking at graphemes might be due to "cross-activation" of V4 (Ramachandran & Hubbard 2001). This cross-activation may arise due to a failure of the normal developmental process of pruning (Ramachandran & Hubbard *The Emergence of the Human Mind: Some Clues from Synesthesia,* **152***)*.

As synesthesia focuses on how the human brain functions, how its left hemisphere can generate poetry under the aegis of specific stimuli and brain areas, it entails the conjunction of poetry, semiotics, neurology and cognitive sciences in general. In this respect, it becomes befitting to provide a few cursory definitions of some technical terms pertaining to these fields all the more as they are invariably used when it comes to analyze and comprehend synesthesia or a synesthetic process: angular gyrus, limbic system, synapses and neurons, psychic areas, brain hemispheres, neo-cortex, and thalamus.

Angular gyrus

It is a region of the brain in the parietal lobe, that lies near the superior edge of the temporal lobe, and immediately posterior to the supramarginal gyrus; it is involved in a number of processes related to language, mathematics and cognition. The angular gyrus is Brodmann area 39 of the human brain, a part of the brain engine rooms (lobes) or more precisely one of the ridges on the surface of the cortex. It is located in a region having the highest level of neural processing, which means: an extremely intense activity of intellectual exchanges is catalyzed by this area especially with respect to cross-modal operations. In *The Emergence of the Human Mind: Some Clues from Synesthesia* scientists such as Ramachandran and Hubbard believe that the angular gyrus is particularly sensitive to the generation of cross-modal thinking and cross-modal metaphors (171-172). It can help truth-seekers to understand cross-modal metaphors and metaphors in general. We will develop and validate this standpoint farther.

Limbic system.

It is a section of the human brain that is responsible for emotion, memory and learning. It has four major components: the hippocampus (memory) and the amygdalae (expression of emotion), the septal area (reward and pleasure) the hypothalamus (control of hormone production, learning, physical reactions associated with emotions such as raised blood pressure, breathing rate due to anxiety, etc . . .). The pre-requisite for the occurrence of synesthesia is an emotional stimulus or inducer, which calls for an activation of the limbic system because emotion induces its activation.

Synapses and neurons

Synapses are points of junction between neurons. **Neurons** are brain cells or nerve cells regarded as physiological and anatomic units of the nervous system. Neurons secrete neurotransmitters, that is, chemicals that carry electrical signals designed for the communication via synapses and neurons in the brain

Psychic areas of the brain

They are areas specializing in the control of specific sensory modes: psycho—visual area (for sight); psycho-auditory area (for hearing); psycho-olfactory area (for smell), psycho-gustatory area (for taste), psycho-tactile area (for touch)

Brain hemispheres

The brain is vertically divided into two main parts called hemispheres: the left and the right. An organ named "corpus callosum" connects the two hemispheres. The left hemisphere controls language, memory, abstract operations and the right side of the body and this explains why the right side of an individual can become paralyzed if the left hemisphere is damaged. On the contrary, the right controls simple operations and the left side of the body. Therefore, both hemispheres function cross-modal wise.

Thalamus

Is located underneath the corpus callosum and processes sensory information on its way to other parts of the brain and is linked with sleep and, possibly, consciousness.

Hypothalamus

A cluster of nuclei located under the thalamus, controls many body functions including feeding, drinking and the release of many hormones.

Neo-cortex

Is acknowledged as the super-brain because it is the most sophisticated part of the brain. It is located in the top layer of both hemispheres. It specializes in highly sophisticated complex operations such as thinking, philosophizing, and problem-solving.

How does synesthesia occur? What is its modus operandi, its scientific aspect?

I—THE GENERATION OF SYNESTHESIA.

The way synesthesia is generated reveals its multidisciplinary status because it predominantly includes the combination of several fields and especially neuroscience, genetics, psycholinguistics, literature, semiotics and, more recently, neuroesthetics.

—MULTIDISCIPLINARY ASPECT OF SYNESTHESIA

A—SCIENTIFIC ASPECT OF SYNESTHESIA

Physiological mechanisms and anatomical explanations

Synesthesia is a brain-generated mechanism assisted by seven sections/organs: 1—the angular gyrus, 2—the limbic system, 3—synapses and neurons, 4—psychic areas, 5—brain hemispheres, 6—neo-cortex, and 7—thalamus. As we mention earlier, it takes place in the neo-cortex and left hemisphere of the human brain because this hemisphere controls language. Precisely, synesthesia involves a highly refined form of language: poetry produced by inter-sensory metaphors. Several researchers: Cytowic, Baron and Cohen have conducted interesting studies on synesthesia. In light of these studies and the current level of research it is possible to substantiate its neurological basis. As a matter of fact, synesthesia is induced by emotion. Specific stimuli spark the activation of the limbic system. Under this activation, emotion is induced. As a result of this, synapses and neurons usually pertaining to two or more different psychic areas of the brain cross to reach others, tangle and merge. This merging triggers the merging of the two or more psychic areas to which these synapses and neurons belonged and this leads to cross-modal connections, that is, synesthesia. Additionally, the whole process is assisted by the thalamus and the angular gyrus because the former (the thalamus) oversees the process of translating mere electrical signals conveyed by synapses and neurons and into concrete information: synesthesia, whereas the latter (angular gyrus) participate in or, at least, hones the production of cross-modal forms of language and abstraction: synesthesia.

It is noteworthy that in the whole process of synesthesia, emotion or emotional response is at the very basis of the mechanism because emotion triggers it and several scientists—and especially Drs Cytowic (*The Man Who Tasted Shapes* Cambridge: MIT Press, 2003), Baron-Cohen (*Synaesthesia, Classic and Contemporary Readings*) agree upon the limbic driven aspect and neurological status of synesthesia since the limbic system controls emotion. Emotion is nurtured by a stimulus or inducer, be it an event, a color, a shape, a sound or whatever, given that there is a multifarious number of them. Let us put things in perspective. When a synesthete S says/feels: "that **noise** is **pink**" what happens inside his/her brain? The psycho-**auditory** area (**noise**) and the psycho-**visual** area (**pink**) merge under the effect of emotion. In this case, synapses and neurons belonging to the psycho-**auditory** area of the brain (**noise**) and those initially pertaining to the psycho-**visual** area (**pink**) tangle and merge (that **noise** is **pink** = auditory modality (noise) + visual modality (pink)), which causes the merging of the psycho-sensory areas to which they belong. Then, by virtue of the psychological law of totality, there occurs a phenomenon of transcendence by means of which only one sensory perception, inter-modal, takes over and emerges. It follows that the initially bimodal perception (2 modalities -> **noise (1)** is **pink (2)**) shifts into one single modal perception but endowed with the two-pronged features of **noise** (1) and **pink** (2) (merging -> 1 modality: **pink noise**). The activity of the thalamus and the angular gyrus finalize the process by assisting in translating the two sensory pieces of information (**noice + pink**) into one (**pink noise**). Consequently, the synesthetic phenomenon comes to fruition. Here, the tangling and merging of synapses are vindicated by the theory of synaptic plasticity and adjacency elaborated by Drs. Ramachandran and Hubbard and mentioned previously (see diagrams and pictures, page 65). This theory stipulates that synapses are flexible, can change in terms of structure and function under certain conditions, which explains their possibility or propensity of tangling and merging especially if they are adjacent. The concept of adjacency (for auditory and visual modalities) is also analyzed and corroborated by Dr. Rampa in Chapter 6 of *You Forever (38-39)* albeit used with different concepts "symbolic keyboard". We will examine it later. It is noteworthy that, in any synesthetic phenomenon, the stimulation is both bimodal (or multimodal) and inter-modal because the stimulation of one sensory mode is simultaneously perceived by the other (s) without any prior stimulation of the latter and only one will transcend all the others and emerge (psychological law of totality). Later on, we will see what occurs in a synesthete's brain through scientific devices utilized in neuroscience and enabling to diagnose the human brain without opening it by means of surgery. One of these devices is Positron Emission Tomography (P. E. T.).

Now, it is interesting to understand how synesthetes perceive reality to glance at the epistemological posture under the aegis of which synesthesia functions. In synesthetes, each letter of the alphabet and each number have a specific color. Dreams are in color. Similarly, each day, week, month and year is inherently colored (color-grapheme synesthesia), has a shape and texture. Most scientists in general and Dr. Cytowic in particular have convincingly elaborated the limbic theory of synesthesia. Precisely, the fact that it is driven by the limbic system and accordingly, emotionally nurtured, accredits this theory. Some believe that synesthesia is genetic and transmitted through the chromosome X. This explains why most synesthetes are women. Indeed, genetics has proven that all women have 46 chromosomes (XX) and men 46 chromosomes (YX). It is clear that in a couple of synesthetes pending a baby girl, the probability for their prospective daughter to be a synesthete is higher than a boy is. Since her *karyotype is 46 XX, that of her mom is 46 XX, her father's is 46 YX, she does not have the chromosome Y. Therefore, as a girl, she can only receive the synesthetic gene, the chromosome X from her mother and the other X from her father. For other researchers everybody has it as a baby and eventually loses or keeps it permanently. In the latter case, the normal pruning of synapses from certain brain regions does not occur, which explains why, these areas remain connected and synesthesia become permanent. That is Ramachandran and Hubbard's theory of permanent synesthesia (op.cit.). Synesthesia involves language, cross-modal thinking and its corollary: a holistic perception of the universe, which means reality is infused with emotion, poetry/artistry, consciousness, memory, feeling(s), and sensation. That is why literature and especially its vicinal areas: linguistics, psycholinguistics, semiotics, as well as neuroscience and psychology, are or should jointly be involved in the study of synesthesia. As a matter of fact, the perception of reality by synesthetes is holistic and meticulous, complex and itemized because they see, hear, feel, sense, perceive or even memorize the world surrounding them with all its minute details. Such a perception is termed "**eidetic**". The Greek root "ΕίΔΟΣ (**eidos**)" means "**essence**". Thus, an eidetic perception is that which systematically provides all the minute details including the very essence of things or facts by giving a very explicit and vivid image of what the observer describes, what he/she perceives or talks about. It is so vivid that it gives the impression to "see" each thing as a three-dimensional image. In *How Synesthetes Color their World*, Patricia Duffy, an expert in the field and synesthete herself substantiate this fact. She says: "What I see is always felt, sensed, and colored. It has a shape, texture, three dimensions, it comes with a specific type of form, mood, sensation, and even a spatial sense. My perception of reality is always, total and vivid." (16, *How Synesthetes Color their World*) Patricia's perspective is confirmed by experts like Baron Simon Cohen, and John E. Harrison in *Synaesthesia, Classical and Contemporary Readings*. Additionally, most synesthetes are endowed with hypermnesia, that is: a highly developed memory. Since they perceive reality holistically, each letter and digit are associated with a specific shape, texture, and color, it becomes much easier for them to memorize things in minute details. The digits they see are consubstantially associated with colors. That is why colors are seen and immediately followed by the corresponding letters. For instance, if 7 is red and 9 is green, in their mind's eye they see the red color immediately followed by green. Thus, red (7) + green (9) = 16. That way, they can memorize a tremendous amount of texts, principles, theorems, and equations quickly and easily. Neurologists like Cytowic in *Synaesthesia A Union of Senses* and Baron Simon Cohen in *Synaesthesia, Classic and Contemporary Studies* give details about hypermnesia. Synesthetes can also hear colors, taste sounds. However, most cases of synesthesia relate to visual-auditory modalities, and a color-grapheme aspect. The manifestation of synesthesia—especially the visual-auditory mode is also clarified by the concept "symbolic keyboard". In lesson six of his valuable book *You Forever*, Dr. Lobsang Rampa mentions what is considered a "symbolic keyboard" and how it relates to synesthesia. According to him, there is a symbolic keyboard presenting a spectrum of sensory perceptions in the universe: sight, sound/hearing, smelling, touch, taste radio. On this keyboard certain sensory perceptions are very close to one another and therefore likely to coalesce, which convincingly correlates with Dr. Ramachadran's concept and theory of adjacency. Sight and sound are adjacent and therefore fall under this category. Based on this fact, it dawns upon us that the close proximity of sight and sound and their easy possibility of blending suggest not only the understanding and the occurrence of bimodal synesthesia through sound and sight (as the commonest aspect of synesthesia), but also the phenomenon of synesthesia in general. In this respect, Dr Rampa states: "You will observe that after sound, we have sight, and there are certain cases in which sounds have been almost seen, 'apperceived' would be a better term, because under certain conditions if a person is clairvoyant he can 'see' the shape of sound. You have probably heard someone say, 'Oh, it was such a ROUND sound' or something similar" (70). It is worth mentioning that the angular gyrus plays an important role in any type of synesthesia.

The importance of the angular gyrus within the framework of synesthesia/cross-modal thinking

The angular gyrus is a brain region of paramount importance in the finalization of synethesia/cross-modal thinking process. Indeed, it is involved in the elaboration of highly sophisticated abstraction and percepts. If it is damaged or dysfunctional, it becomes impossible to name things, make abstract connections or accomplish any cross-modal synthesis. Ramachandran and Hubbard provide a compelling analysis of its role in the section of their article titled *The Angular Gyrus, Multisensory Convergence, and the Evolution of Abstraction*. They state:

> Depending on the stage at which cross-activation occurs, there might be "higher synesthetes" and "lower synesthetes" and that in the former the cross-talk might be in the general vicinity of the angular gyrus, which is known to be involved in higher numerical concepts. The idea that some types of synesthesia might involve the angular gyrus is also consistent with the older clinical observation that this structure is involved in cross-modal synthesis; information from touch, hearing, and vision is thought to flow together in the angular gyrus to enable construction of high-level percepts. For example a cat is fluffy (touch) and it meows (hearing) and has a certain appearance (vision) and odor (smell), all of which are evoked by the memory of a cat or the sound of the word "cat". No wonder patients with damage to the angular gyrus lose the ability to name things (anomia) even though they can recognize them. Additionally, these patients have the

difficulty with arithmetic, which also involves cross-modal integration; you learn to count with your fingers. (Indeed, if you touch the patient's finger and ask him which one it is, he often cannot tell you.). All of these bits of clinical evidence strongly suggest that the angular gyrus is a great center for cross-modal synthesis. So perhaps it is not so outlandish after all that a flaw in the circuitry could lead to colors quite literally evoked by certain sounds. We have noticed that patients with anomia caused by lesions of the left angular gyrus also have difficulty with metaphors. [. . .] Could it be that the angular gyrus, which is disproportionately larger in human compared to apes and monkeys, originally evolved for cross-modal abstraction, but then became co-opted for other types of abstraction such as metaphors as well? (It would be interesting to see, though, whether these patients are even worse at cross-modal metaphors such "loud shirt" than at other types of metaphors.) (173)

However, how does synesthesia take place in the brain? During a synesthetic process including an audio-visual modality for instance, the limbic system is activated, which sparks an intensive activity in the psycho-visual and psycho-auditory areas and generates a substantial amount of brainwaves that impact the optical and acoustic nerves. Then, the waves appear under the form of either light or sound (because they are binary by nature and they can appear under the form of light or sound). However, the psychic areas (psycho-visual and psycho-auditory areas) responsible for discriminating them into light or sound merge and this results in synesthesia. The whole phenomenon is assisted by the left hemisphere, the neo-cortical encephalic area, the limbic system, the angular gyrus, and the thalamus. It therefore dawns upon us that the discrimination of waves into light and sound does not occur. Similarly, the discrimination of visual modality and auditory modality does not occur in the synesthesia phenomenon. However, to fully appreciate synesthesia as a factual phenomenon, it is necessary to prove it scientifically.

Scientific evidence of synesthesia: two major cases

1—Case of synesthesia verified by P.E.T (Positron Emission Tomography)

In order to scientifically assess the reality of synesthesia, specialists use many functional neuroimaging and revolutionary techniques. One of them is a special device called **Positron EmissionTomography** (P.E.T.). It is a powerful scientific investigation tool designed to scan the brain and assess the quantity and the quality of blood flow in it. It can also be utilized to scan the brain in order to detect any possible malfunction. Thus, by means of P.E.T. researchers have scientifically verified the existence of synesthesia. As a matter of fact, when P.E.T. is activated, a tremendous amount of blood flows in the **psycho-visual area and the psycho-auditory area** of the synesthete's brain **simultaneously** (in case of bimodal mode of synesthesia with **sight and hearing**. For instance: "he **heard a gorgeous color**"). On the contrary in a non synesthete's brain, during P.E.T. activation the blood flows **either** in the **psycho-viusal area** of the brain or in **the psycho-auditory area** only. All this clearly appears on the P.E.T. screen and attests to the scientific evidence of synesthesia.

Functional neuroimaging studies using positron emission tomography (PET) and functional magnetic resonance imaging (fMRI) have demonstrated significant differences between the brains of synesthetes and non-synesthetes. The first such study used PET to demonstrate that some regions of the visual cortex (but not V4) were more active when auditory word -> color synesthetes listened to words compared to tones (Paulesu et al. 1995). More recent studies using fMRI have demonstrated that V4 is more active in both word -> color and grapheme -> color synesthetes (Nunn et al. 2002; Hubbard et al. 2005a; Sperling et al. 2006). However, these neuroimaging studies do not have the spatial and temporal resolution to distinguish between the pruning and disinhibited feedback theories. Future research will continue to examine these questions using not only fMRI but also diffusion tensor imaging (DTI), which allows researchers to directly investigate neural connectivity in the human brain and magnetic resonance spectroscopy (MRS) which allows researchers to measure the amounts of different neurotransmitters in the brain.

THE FOLLOWING IS AN IDEA OF WHAT P.E.T. IS:

Description

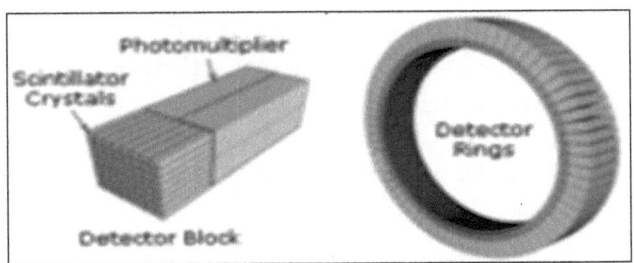

Schematic view of a detector block and ring of a PET scanner
(here: Siemens ECAT Exact HR+)

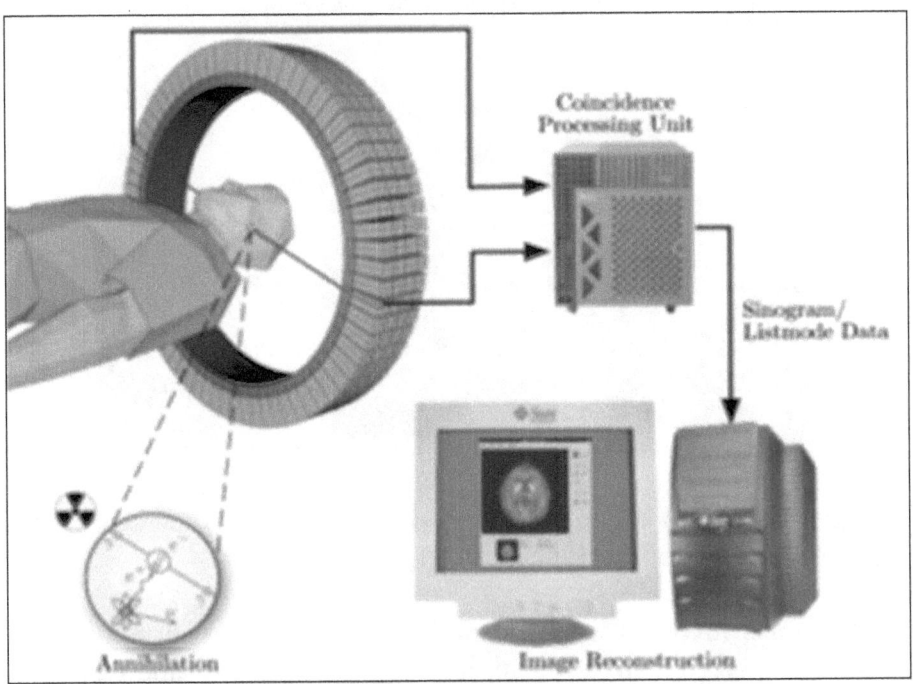

Schema of a PET acquisition process

How can we measure brain activity?

The brain cells communicate by sending tiny electric signals to each other. The more signals that are sent, the more electricity the brain will produce. An EEG can measure the pattern of this electrical activity. Active areas of the brain also use more energy than less active parts—this is the basis of PET and fMRI scanning. In a synesthete's brain a substantial amount of blood simultaneously flows in specialized psychic areas, which is not the case for a non synethete's brain. For instance in the case of an visual auditory mode like: a red (visual) sound (auditory), a substantial amount of blood simultaneously flows in the psycho-visual area and audio-visual area.

THE FOLLOWING IS A SYNESTHE'S BRAIN DURING P.E.T. ACTIVATION:

Cross-activation

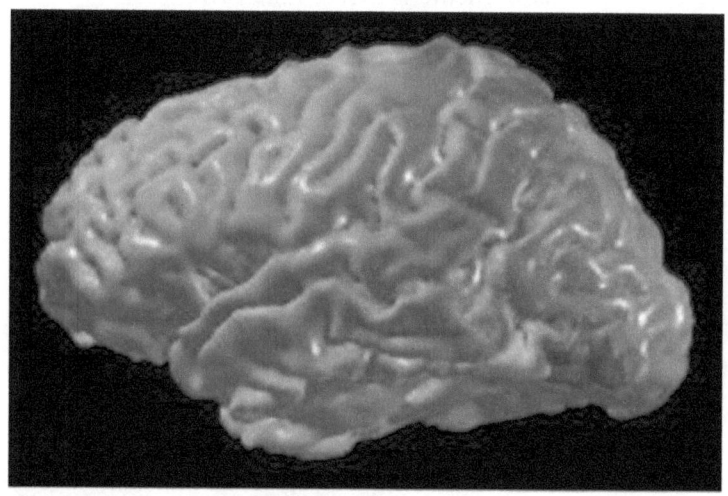

Regions thought to be cross-activated in grapheme-color synesthesia (from Ramachandran & Hubbard 2001).

Since regions involved in the identification of letters and numbers lie adjacent to a region involved in color processing (V4), the additional experience of seeing colors when looking at graphemes might be due to "cross-activation" of V4 (Ramachandran & Hubbard 2001). This cross-activation may arise due to a failure of the normal developmental process of pruning, which is one of the key mechanisms of synaptic plasticity, in which connections between brain regions are partially eliminated with development. Similarly, lexical -> gustatory synesthesia may be due to increased connectivity between adjacent regions of the insula in the depths of the lateral sulcus involved in taste processing that lie adjacent to temporal lobe regions involved in auditory processing. Similarly, taste -> touch synesthesia may arise from connections between gustatory regions and regions of the somatosensory system involved in processing touch. However, not all forms of synesthesia are easily explained by adjacency.

2—Case of color-grapheme synesthesia verified by iterative written reports

The verification of this type of synesthesia is based on reports conducted on letters and numbers and their corresponding colors seen by synesthetes. Letters and numbers are inherently colored. These experiments were repeated many times on the same subjects and every single time specialists conducted them they reported the same results with the same synesthetes: the same letters and numbers consistently matched the same colors, shapes and textures. Such tests, repeated plenty of times scientifically certified the reality of synesthesia.

Grapheme -> color synesthesia The following is how a synesthete might perceive certain letters and numbers

In one of the most common forms of synesthesia, grapheme -> color synesthesia, individual letters of the alphabet and numbers (collectively referred to as graphemes), are "shaded" or "tinged" with a color. While synesthetes do not, in general, report the same colors for all letters and numbers, studies of large numbers of synesthetes find that there are some commonalities across letters (e.g., A is likely to be red) (Day 2005; Simner et al. 2005).

A grapheme -> color synesthete reports, "I often associate letters and numbers with colors. Every digit and every letter has a color associated with it in my head. Sometimes, when letters are written boldly on a piece of paper, they will briefly appear to be that color if I'm not focusing on it. Some examples: 'S' is red, 'H' is orange, 'C' is yellow, 'J' is yellow-green, 'G' is green, 'E' is blue, 'X' is purple, 'I' is pale yellow, '2' is tan, '1' is white. If I write SHCJGEX it registers as a rainbow when I read over it, as does ABCPDEF."[2]

Another reports a similar experience. "When people ask me about the sensation, they might ask, 'so when you look at a page of text, it's a rainbow of color?' It isn't exactly like that for me. When I read words, about five words around the exact one I'm reading are in color. It's also the only way I can spell. I remember in elementary school remembering how to spell the word 'priority' because the color scheme, in general, was darker than many other words. I would know that an 'e' was out of place in that word because e's were yellow and didn't fit.

Another reports a slightly different experience. "When I actually look at words on a page, The letters themselves are not colored, but instead in my mind they all have a color that goes along with them, and it has always been this way. I remember back in kindergarten thinking that each homeroom had a different color associated with it. I would sometimes say things referring to that class and calling it by its color. It is also like this with days of the week, months, and so on." I thought this was caused by me over-thinking things. "But I finally have come to realize that Synesthesia is real." The following image provides an idea on how a synesthete sees months.

ANISOTROPY IN A SYNESTHETE COMPARED TO A NON SYNESTHETE

Anisotropy is the property of being directionally dependent, as opposed to **isotropy**, which means homogeneity in all directions. It can be defined as a difference in a physical property (absorbance, refractive index, density, etc.) for some material when measured along different axes. An example is the light coming through a polarising lens. In most synesthetes one can find cases of anisotropy rather than isotropy, which includes the possibility of hearing colors when watching pictures but not the possibility of watching pictures to hear colors. Therefore, their synesthetic faculty operates in one sole direction because these synesthetes are directionally dependent. However, it is possible to find a modicum of cases of isotropy among synesthetes. That is what Van Campen calls "two way traffic" in *Synesthesia, the Hidden Sense.(16)*

Richard Cytowic has also done a remarkable work on synesthesia. Still, it is noteworthy that it is a relatively new field and any research and finding are provisional and should be completed, amended or readjusted as new discoveries will be available over time. In addition, for the sake of efficiency, researchers should work in multidisciplinary teams or at least seek support from other specialists also interested in synesthesia: semioticians, linguists, psycholinguists, poets, psychologists, etc . . . because it is a jurisdiction where "intellectual democracy" should be exercised.

MY PERSONAL THEORY ON UNDERSTANDING SYNESTHESIA IN LIGHT OF A MODICUM OF PHYSICS

Given that I use a multidisciplinary approach (even if my main corpus is literature), and that synesthesia is at the core of most fields of human knowledge, it is possible to use physics, just like many other disciplines, to explain and understand it. De facto, nowadays, on the basis on modern physics heralded by Einstein, we know that matter and energy are inter-convertible and, accordingly, interrelated: $E = M.C^2$, where **E** stands for **energy**, **M** for **mass/matter** and **C** for **celerity**, that is: the speed of light (300,000 kilometers per secundum). From this equation it can be inferred that if we accelerate mass/matter at an extremely high speed, near that of light, it (even if it is a very tiny portion) will generate a tremendous amount of energy capable of destroying a whole city. That was the key-idea "feeding" the atom bomb. The destruction of Hiroshima and Nagasaki is a tragic and patent fact of such power. Just like energy and mass/matter are interrelated, light (sight) and sound (hearing) are interrelated. Indeed, both light and sound/electromagnetic radiation can be viewed as waves or as particles of energy. As a wave in an abstract electromagnetic field or as a stream of massless particles/ corpuscles called "photons" (realm of quantum physics). For our purpose in terms of synesthetic phenonemon, we will opt for the wave description, more pertinent here. Like light, sound can also be regarded as a wave or energy since both share a number of common properties. As waves, they are endowed with **energy** and their main difference is in terms of frequency.

The frequency and wavelength can be determined by the formulae:

$$\text{frequency} = \frac{c}{\lambda}$$

$$\lambda = \frac{c}{\text{frequency}}$$

where 'c' is celerity, the speed of light in centimeters per second, the Greek letter lambda λ is the wavelength in centimeters and the frequency (measured in hertz) is in cycles per second

Also, the photon energy can be calculated by the formulae:

$$e = h \cdot f$$

$$e = \frac{c \cdot h}{\lambda}$$

where 'e' is energy (ergs), 'f' is frequency (cycles per second), 'h' is Planck's constant (4.13566 x 10^{-15} electron volts/second) and wavelength λ is in centimeters.

Thus, during a synesthetic process, the brain is activated, which sparks an intensive dynamism and generates a substantial amount of brainwaves that impact the optical and acoustic nerves. Then, the waves appear under the form of either light or sound (because they are binary by nature and they can appear under the form of light or sound). However, the psychic areas (psycho-visual and psycho-auditory areas) responsible for discriminating them into light or sound merge because of the cross-wiring process resulting from the tangling and merging of synapses and their specific psychic areas (psycho-visual and psycho-auditory areas), merging catalyzed by emotional stimuli (limbic system). Consequently, the brain area designed to process information (the thalamus) does not discriminate light waves (sight) and sound waves (hearing). It follows that synesthesia occurs since sight and sound merge. That is my explanation of synesthesia based on elementary physics. Now it is important to study its literary and artistic aspect.

B—LITERARY AND ARTISTIC ASPECT OF SYNESTHESIA

Synesthesia is a special set of metaphors, cross-modal metaphors. They are special in the sense that they are highly refined, polysemic, endowed with a high coefficient of poeticity, nurtured by brain structures: the angular gyrus, the left hemisphere, neocortex, and driven by the limbic system. All these features make them an eminently subtle and vivid form of literary expression. An analysis of synesthesia from a literary perspective can enlighten us on their modus operandi.

Analysis of synesthesia from a literary perspective

Synesthesia sparks a sophisticated reflection on the cognitive process pertaining to specific brain areas in terms of cross-modal thinking/ abstraction or cross-modal metaphors. The phenomenological analysis of synesthesia stems from the faculty of making connections/analogies between seemingly unrelated areas (sight and hearing for instance) and leads to a better understanding of certain psychic structures of the brain. It appears that the production of cross-modal thinking and metaphors pre-exist in high-level percepts anchored in specific brain locations. Ramachandran and Hubbard elucidate this important viewpoint:

> Numerous high-level concepts are probably anchored in specific brainregions or maps and, according to experts such as Ramachandran and Hubbard, they are represented in a small brain region called the "angular gyrus." For these experts, the mutation that brings about synesthesia causes excess communication between brain regions and maps representing specific perceptual entities such as roundness or sharpness of shapes. Depending on where and how widely in the brain the trait was expressed, it could lead to both synesthesia and to a propensity to link seemingly unrelated ideas; that is, creativity as found in poetry/metaphor (152).

From this perspetive, it appears that synesthesia proves to be a highly refined mode of literary expression. It gives an idea on how the human brain, under certain conditions, generates poetry from the left hemisphere and the neo-cortex and by means of the limbic system and the angular gyrus. It also leads to the understanding of the philosophical function of synesthesia. With respect to the neurological framework of synesthesia, it elucidates the creativity of poets. The propensity and the easiness to use metaphors (the most poetic figure/stylistic device) and *tropes by artists and poets (synesthete poets in particular) suggest that their brains are set up to make connections/analogies between seemingly unrelated domains **(different sensory perceptions)** just as metaphors are conceived to make connections/analogies between seemingly unrelated realms. To appreciate the complexity and refinement of synesthesia as metaphors in its fullness, it is important to explore its morphological components.

MORPHOLOGICAL COMPONENTS OF SYNESTHESIA

Synesthesia **entails:**

- The combination of two or more sensory modalities;
- A copula usually linking these modalities;
- The adjunction of a holistic perception of reality, which is like an extra sense, a spatial one, conceived to endow synesthetic signs

with implicit symbolism and, by the same token, maximize their semantic power.

a—The combination or synthesis of sensory modalities

Synesthesia is conditioned by the combination of two or more sensory modalities endowed with a cross-modal-wise status, which makes synesthesia a cross-modal metaphor also called inter-sensory metaphor because it crosses and associates several sensory perceptions. The process generated by synesthesia systematizes the conjunction of two or more seemingly unrelated domains: visual, auditory, tactile, olfactory, and gustatory. Since synesthesia results from the activation or stimulation of the limbic system (triggered by emotion), this stimulation is both **bi**modal (since **two** sensory modes merge)/**multi**modal (**several** modes merge) and **inter**modal (achieved between different modes). In other terms, the stimulation of two or more inter-sensory modes automatically induces that of all the others. However, when all the modes merge, only one of them transcends all the others, and emerges by virtue of the psychological law of totality. By emerging it intrinsically bears **all** the characteristics and semantic features of **all the others**. Consequently, its meaning becomes the mathematical product of all the others, hence the semantic density of synesthesia. e.g. He was wearing a **blue yelling** shirt. In this instance we have the combination of two sensory modalities: visual (**blue**) and auditory (**yelling**). This combination is nurtured by a high semantic density because it entails not only the conjunction of two modalities but, also and more importantly, the product of all the semantic features relating to the visual and auditory, their underlying semantic ramifications and implied symbolism.

b—The copula/the collage

The word "copula" stems from the Latin verb "copulare" meaning to bind, to connect, to link, to tie and the word "copula, copulae": a tie, a bond. In light of this etymology, we realize that the copula stands for the collage. It is a linguistic connector, a small element used to connect two or more elements, words, or segments within a sentence or several sentences. In a synesthetic construction it is used to finalize the conjunction of two or several sensory modalities or two or more linguistic or extra-linguistic items. Therefore, it performs the synthesis and the transcendence occurring within the process of a synesthetic metaphor. The copula can be presented as a preposition (of), conjunction (and), verb, adjective, or consubstantially be formed with a verb or noun. In this case, it exists under a latent form (blue **yelling** shirt -> here the copula is consubstantial with "yelling", "built in" the verbal form "yelling"). Otherwise, it exists outside the synesthetic signs. e.g. A smoke **of** sounds.

> The copula (—whether it is prepositional, conjunctional, verbal, adjectival, or latent—) is conceived to achieve the collage by finalizing the synthesis within the synesthetic metaphor and mediate the application of the law of totality, which vouches for transcendence. Without copula, there is no synthesis and, accordingly, no synesthesia.

c—The holistic perception of reality/the gestalt vision of the cosmos

Synesthesia always occurs holistically, that is, by means of all the cognitive faculties: memory, reflection, mood, emotion, sensation, feeling, shape of things, numbers inherently colored, dimensions, etc . . . In other words, a synesthete senses, sees, hears, touches things and experiences them concomitantly with the shapes, textures, feelings and sensations they evoke, symbolize, suggest, and elicit. Reality is presented to him/her with **all** its multifarious aspects. That is why all its features are expressed in the metaphorical synesthesia **as a whole**. Therefore, it provides a *gestalt image of this process. The word "gestalt" borrowed from German means total, global, overall. It is specially chosen to underscore this meaningful perception of entirety that always accompanies a synesthetic metaphor. De facto, meaning is exponentially added because it includes two seminal factors: 1—the combination of several sensory modalities 2—the global perception of reality nurtured by mood, emotion, feeling, sensation (s), shape, texture, color(s) and their tints, digits, space, and dimensions (vividness of things seen in three dimensions). It follows that semes, elements of meaning simultaneously feed all the items of this complex spectrum and apply to all its combined items. That is why, the metaphorical synesthesia has a very rich, prolific *semanticism. By semanticism, we mean the extent or range of meaning(s) within a definite text or graphic system as explained in semiotics.

In the instance: "a cold silence", we have a bimodal synesthesia obtained from the coalition of a tactile sensory perception (cold) and an auditory sensory perception (silence) + the holistic perception inherently associated with this metaphorical synesthesia: mood, emotion, feeling, sensation (s), shape, texture, color(s) and tints, digits, space, and dimensions. In this synesthesia, we have not only the coalition of two modal perceptions but also the translation of meaning expressed by the speaker regarding his mood, emotion, feelings, sensation, the shape, texture, colors of what he/she hears and all the underlying suggestions in terms of symbolism at this particular time. Mindful of these considerations, we are prone to consider synesthesia as a unique case of metaphor. Synesthetic metaphors can be represented by the following formula:

Met. Syn -> Σ (2 or more sensory modalities) + gestalt semes inherent in the holistic perception of reality

"Met. Syn" stands for "Metaphorical synesthesia", "->" stands for "is re-written"

Metaphorical Synesthesia is re-written: Product of Two or More Sensory Modalities + Gestalt Semes, that is minimal distinctive units of meaning inherent in the holistic perception of reality.

Given that metaphorical synesthesia can be better understood if it is compared to classical metaphor, it is convenient to explore the differences between classical metaphors and synesthetic metaphors

Classical metaphors and synesthetic metaphors

Classical metaphors and synesthetic metaphors are based on the principle of analogy or connection, as we mention earlier, between seemingly unrelated areas. However, the metaphorical synesthesia is richer and much more refined.

The classical metaphors

The word "metaphor" is composed of the Greek prefix "meta" which means "beyond", "on the other side of", "across", and the verb "phorein": to carry. In light of this etymological breakdown, we can define the metaphor as a stylistic device in which one transfers the literal meaning of a word to another by virtue of an implicit comparison or analogy. The metaphor generates a semantic shift from one item of a specific nature to another. Such a shift highlights and superimposes all the underlying meanings or elements of meaning from each item and confers upon the metaphor a highly refined semantic power. Unlike a simile, a metaphor does not employ conjunctions such as "like", "as". Thus, it appears that comparison is explicit in a simile and implicit in a metaphor. The metaphor is usually viewed as a poetic figure because it is endowed with the sophisticated ability to create meaning(s), refine it (them), proliferate it (them). At its most refined level, the metaphor culminates into a symbol. Consequently, a symbol can be a kind of metaphor that has reached such a high level of esthetic refinement that it epitomizes something. At that stage, it becomes an archetype, a perfect model conceived to represent a quality or a deficiency. For these reasons, it is often used by poets and even regarded as their creative instrument par excellence. That is why in *Lectures Sémiotiques*, Roland Barthes refers to it as "the daily bread of poets". As mentioned earlier, the metaphor is designed to generate and highlight meaning (s) by creating an analogy and special connections (between two or more items of seemingly different nature) that do not generally exist or are difficult to make in prose. Establishing such connections presupposes three data:

- **First**, the possibility of shifting, "sliding" from one meaning (literal) to another (symbolic/allegorical, etc . . .) by complying with the semiotic principle of semantic indirection that we studied in Chapter II. According to it, language can be deceptive in such a way that a word or a group of words may say/express something to mean something else.
- **Second**, the possibility of generating isotopies. The notion of **isotopy* refers to the repetition of a basic meaning trait (seme); such repetition, establishing some level of familiarity within a text, allows for a uniform reading/interpretation of it. An example of a sentence containing an isotopy is *I drink some*

water. The two words *drink* and *water* share a seme (unit of meaning and a reference to liquids), and this gives homogeneity to the sentence. This concept, introduced by Greimas in 1966, had a major impact on the field of semiotics, and was redefined multiple times. Catherine Kerbrat-Orecchioni extended the concept to denote the repetition of not only semes, but also other semiotic units (like phonemes for isotopies as rhymes, rhythm for prosody, etc.). Umberto Eco showed the flaws of using the concept of "repetition", and replaced it with the concept of "direction", redefining isotopy as "the direction taken by an interpretation of the text".

However, here we will adopt a definition of isotopy on the basis of Greimas's and Eco's standpoints that is: the direction of interpretation and category of meaning that is deployed, extended throughout a text. Such a direction and meaning are prone to be created by the mechanism of the metaphor.

- **Third**, the creation of poetic shocks out of the semantic shift and the superimposition/product of all the underlying meanings nurtured by the device. Poetic shocks stem from the semantic distance between the items involved in the analogy. The farther the distance, the more powerful the meanings become, the greater the shocks are, and the more poetic the metaphor/text becomes. All this process confers upon the metaphor a highly refined semantic power. **e. g. This lady is a flower**.

In this example, a flower literally designates a plant and a lady a human being of female gender. However, a human being cannot be a flower stricto sensu. It turns out that, in this instance, the word "flower" is used to evoke, suggest beauty because, through its beautiful and appealing aspect, it has come to generally symbolize the concept of beauty in many languages and cultures. Thus, the literal meaning (**human being**) is—by semantic indirection—transferred to another meaning, figurative (**flower -> beautiful person -> the flower suggests beauty**) by virtue of an implicit analogy/comparison/connection: **sharing a common quality: beauty -> the lady 'houses' beauty, the flower 'houses' beauty as well** (she is **beautiful like a flower**). With respect to the principle of isotopy, it can be generated by creating an internal distension of the metaphorical discourse, which factually leads to the emergence of two isotopies:

- The lady **is not** a flower = plant since she **is a human being, isotopy 1**;
- The lady **is** a flower = gorgeous, very beautiful because the word "flower" suggests, evokes, or symbolizes the beauty, charm, and attractiveness pertaining to a flower. Since the intrinsic qualities of the flower (beauty, attractiveness, glamour, etc . . .) are transferred to her, she becomes **a flower (by analogy), isotopy 2**

Human Being/Female = first isotopy -> seme of human species

The lady

Flower = second isotopy -> seme of vegetal nature

The semantic shift enables the discourse to slide from the literal meaning (human being) to the figurative meaning (flower) and the semantic distance between them simultaneously generates an internal distension of the metaphorical discourse and a poetic shock.

Those are the major characteristics of the classical metaphor. How about the synesthetic metaphor?

The synesthetic metaphor

It demarks itself from the classical on the basis of several esthetic canons: **epistemological complexity and poetic quality; idiolectic pertinence; three levels of meaning**

1—Epistemological complexity and poetic quality.

The metaphorical synesthesia is a special case of metaphor. De facto, due to its epistemological and cognitive complexity, it needs to be explained through the diligent combination of neuroscience and literature/semiotics because any attempt to analyze it through the narrow prism of either literature/semiotics alone or neuroscience is doomed to provide shallow information or even leads to mere failure. Moreover, there is something eminently striking and interesting in the metaphorical synesthesia: its poetic quality. As a matter of fact, synesthesia is predisposed to be poetic. It indeed provides a pertinent analysis of how poetical structures are embedded in the human brain under the form of cross-modal anatomical maps located in the angular gyrus. These structures are "catalyzed" by emotion which is under the control of the limbic system. As we have studied earlier, the limbic system is the part of the brain controlling learning, memory and especially emotion. Under the aegis of emotion, the limbic system is activated. As a result of this, synapses and neurons initially pertaining to two different areas of the brain tangle/merge and this merging causes the merging of the two psychic areas to which they originally belonged. The psychological law of totality synthesizes the different sensory modalities into one and its process culminates into synesthesia. It follows that emotion is at the very core of poetic activity. It is also one of its pre-requisites. Indeed, it nurtures inspiration and creative imagination. Actually, the word "poetry" derives from the Greek verb "poeien" ("ποειν") which means "to create". In light of these considerations, we see a clear correlation between emotion, inspiration and poetic creation (imagination). These three concepts are so mutually inclusive that they constitute a subtle triptych. Plato validates them in *The Republic*. It is indeed noteworthy that in the *Seventh Book of The Republic* he reflects upon poets and poetry. He believes that emotion impacts imagination, frees it and cogently assigns it with the function of creating. At the end of his analysis, he comes to realize that lovers are potentially poets because they are very emotional, which conditions them to create with their imagination and, accordingly, acquire poetic skills. Precisely, since metaphorical synesthesia is engineered by emotion/ the limbic system, it definitely becomes endowed with a poetic quality and, by the same token, it reveals a tacit causality principle: emotion (induces inspiration) -> imagination -> creation. In his article section titled "*Why Is Synesthesia More Common in Artists, Poets, Novelists, and Composers?*" Ramachandran and Hubbard gives an insightful explanation regarding this important point. He outlines:

> Do artists and poets seem to have synesthesia because they have a vivid imagination and they are prone to talk as if they do? Or is their creativity linked in some deeper way to their synesthesia? Is synesthesia the result of a vivid imagination?

Or is a vivid imagination the result of synesthesia? [. . .] One thing that artists, poets, and novelists have in common is that they are specially good at using metaphor ("It is the East and Juliet is the Sun"). It is as if their brains are set up to make links between seemingly unrelated domains, like the sun and his beautiful young woman [. . .] Just as synesthesia involves making arbitrary links between seemingly unrelated perceptual entities like colors and numbers, metaphor involves making links between seemingly unrelated conceptual realms. Perhaps this is not a mere coincidence. Perhaps the reported higher incidence of synesthesia in artists is rooted deep in the architecture of their brains. (169-170)

Numerous high-level concepts, percepts are probably anchored in specific brain regions or maps and, they are represented in a small brain region called the "angular gyrus." For these experts, the mutation that brings about synesthesia causes excess communication between brain regions and maps representing specific perceptual entities such as roundness or sharpness of shapes. Depending on where and how widely in the brain the trait was expressed, it could lead to both synesthesia and to a propensity to link seemingly unrelated ideas; that is, creativity as found in poetry/metaphor. (152)

All this process can be represented by the following diagram:

DIAGRAM OF POETIC CREATION INHERENT IN SYNESTHESIA

=> 1—Brain structures: angular gyrus <-Limbic system -> emotion/inducer (stimulus)
=> 2—Tangling of synapses belonging to different psychic areas
=> 3—Merging of their psychic areas => Synesthesia => Freed Imagination => Creation => Metaphor/Synesthesia/Poetry

In addition to its epistemological complexity/poetic quality, the metaphorical synesthesia stands out through its idiolectic pertinence.

2—Idiolectic pertinence

Idiolectic pertinence is opposed to classematic pertinence. The synesthestic metaphor possesses idiolectic pertinence whereas the classical metaphor has a classematic one. The former is elucidated by the fact that synesthesia can seemingly be outlandish, weird, strange, but in spite of its weirdness it is so powerfully and artistically constructed that it becomes convincing from the point of view of its signified (s) and signifier. Therefore, it intrinsically contains outstanding qualities whereby it asserts its inner truth as a corpus of poetic signs. Idiolectic pertinence can be assessed by two criteria: the power of suggestion and latent symbolism.

A—Power of suggestion

The power of suggestion is expressed by a meticulous conjunction of words on the basis of their remarkable semantic effects and their ability to nurture underlying meanings appending symbolic implications.

B—Latent symbolism

Synesthesia usually occurs with latent symbolism because the collusion and collision of its sensory modalities have the propensity to produce meaning multiplied by their semantic field. Each word is surrounded by a semantic field (semantic field theory), which provides it with the possibility of generating meaning. Semiotic studies show that this meaning vividly increases with the effect of the collage to suggest additional semantic ramifications, symbolic by nature. When symbolic meaning prevails in a text, it also generates other symbols and, by the same token, other meanings. Such semiotic posturing gives way to the semantic prolificacy inherent in a symbol as it has been scrutinized by semioticians on the basis of its ability to reproduce meaning and many levels of meaning. In this area, Genette, Roland Barthes and Krysteva have conducted pertinent studies. Their compelling analysis has proven that the meaning of a symbol can never be exhausted and its reading therefore becomes a manipulation. Latent symbolism adheres to the hermeneutic principle of non-exhaustiveness status of the sign as it was expounded in *Figures II* by Gerard Genette.

Examples

a) Her smile is singing.

Example a highlights the special beauty of the smile and, by symbolism, that of the lady who smiles. This metaphor has a special effect validated and generated by the conjunction (or the collusion/collision) of two sensory modalities meticulously chosen/presented: smile (visual) and singing (auditory). Accordingly, they cause the metaphor to stand out by means of its power of suggestion and latent symbolism which, definitely, substantiate its pertinence. Consequently, in spite of its unusualness and even weirdness, this metaphor is intrinsically convincing. It follows that its pertinence is idiolectic.

b) Her smile's song is a signature of God

Example b assumes the same semiotic features, but it carries them a step further because of the adjunction of another metaphor ("signature of God") to the first ("smile's song"). This added metaphor is eminently suggestive because of the refined way in which it is presented (its sophisticated signifier), the meticulous choice of words used, and its tacit symbolism (its signified): "God's signature". All these characteristics make it stand out sub specie aeternitatis. Finally, it becomes a real masterpiece. Consequently, its pertinence is more idiolectic than that of the first.

c) This man is a lion

Example c shows a classical form of metaphor. As a matter of fact, the word "lion" is commonly and universally used to symbolize/suggest power, strength, and majesty. Indeed, this animal is regarded as the lord of the forests because of its physical power, exceptional charisma, majesty, and kingliness. Thus, by implicitly comparing a man to a lion, one confers upon him the amazing qualities universally associated with a lion. It appears that, depicting a man as a lion is conceived to valorize his outstanding strength and majesty. However, this metaphor has been so extensively used that it has almost become a case of jurisprudence, a kind of law, and a classical instance. Consequently, its pertinence is said to be classematic inasmuch as one can expect a man to be granted these "over-mentioned" qualities. Conversely, one can neither expect "a smile to sing" nor its "being a signature of God". In the final analysis, these instances show that idiolectic pertinence is governed by the **unexpected, the esthetics of the surprise** whereas the classematic is regulated by the **expected, the esthetics of the non-surprise.**

Besides, the synesthetic metaphor is so vivid and complex that it has three major levels of meaning.

3—Three levels of meaning: literal/figurative, holistic, ontogenetic

a—literal/figurative level

Unlike the classical metaphor in which the literal meaning is strategically and temporarily "discarded" to highlight the figurative by semantic indirection, the synesthetic metaphor takes the literal meaning into account. Indeed, what the poet/synesthete perceives is factually real and literal even if it is formulated in a flowery and figurative language. In spite of the fact that cross-modal metaphors are unusual, morphologically complex and sophisticated, their meaning is literal **and** figurative. Given that what they refer to is usually complex, suggestive, and infused with underlying symbolic ramifications, they can be both **literal and figurative**. Therefore, they become a unique case of metaphors whose semiotic structure reveals the superimposition of two semantic sub-levels: literal and figurative.

b—holistic level

The synesthetic metaphor always occurs with a global perception of reality. De facto, through it, reality appears in all its multifarious aspects because it comes holistically, as a total corpus (a group of rich elements) and not monistically (as one single element). The word "gestalt" borrowed from German pertinently expresses such a perception, that is: "global, total, as a monolithic unit". Perception is therefore manifested in a gestalt aspect: through feeling, sensations, letters and numbers inherently colored, texture, form, emotions and a psychological disposition consubstantially associated with all the features of the bi-modal or multi-modal sensory combination. As an illustration, in the example "a shouting green" the metaphor not only suggests the combination of two sensory modes and their implied combined meanings but also the manifestation and the meanings of the holistic perception revealed through it (meanings affecting feeling, numbers, sensations, letters inherently colored, texture, form, emotions and the consecutive psychological disposition.) To put it in a more simple way, we can consider that, in "a shouting green", the speaker not only sees/hears the green but also suggests what he feels, senses: the texture, shape, three dimensions of the perception, the tint of the colors inherent in all this auditory-visual complex (shouting green) as well as the mood in which he is when he sees/hears the green or the effect that the audiovisual complex produces on him. From this perspective, it can be inferred that there are two types of semes: those appearing at the first level or **proto-semes** of the metaphors (from Greek **'protos' = first**) and those that are suggested, latent or **crypto-semes** (from Greek kruptos, *hidden*, kruptein, *to hide*). Their expression occurs at the second level and needs to be analyzed to be understood by non synesthetes or/and monitored by scientific devices (P.E.T., M.R.I.) designed to explore the synesthete's brain. So, from a semantic viewpoint, the synesthetic metaphor is much more refined and richer than the classical.

c—ontogenetic level

The ontogenetic level is a corollary, a logic sequel of the holistic level. Synesthesia complies with an ontogenetic process. The word "ontogenetic" stems from the Greek "onto", "ὤν" ("on"), ὄντος ("ontos") "being; that which is", present participle of the verb εἰμί "be", and—λογία,—logia: science, study, theory. Regarding "genesis", it refers to "origin, creation, production/reproduction". Thus, this notion refers to the ability for synesthesia to generate or reproduce meaning on its own and continue to generate it exponentially. In the final analysis, meaning almost becomes an endless activity. In *Semantics of Poetry*, Barthes and Cohen substantiate this ability for specific forms of language to generate meaning endlessly under certain conditions. Performing this task presupposes that metaphorical synesthesia is in itself endowed with semiotic qualities that can be utilized and developed for that purpose. So, what are these intrinsic qualities? The semantic field theory and the very status of the symbol provide the answer. The semantic field theory posits that each word is surrounded by a field of meaning and it is the context in which words are used that enables us to decipher their meaning(s). Pragmatically, the field of meaning can significantly increase with lexical vicinity, that is the proximity of other words and their mutual *syntagmatic and *paradigmatic interactions. Such proximity acts like a magnetic field that systematically attracts meanings, superimposes and multiplies them. With respect to the status of the symbol, it shows that it is endowed with the capacity (or the potential ability) of generating meaning constantly under appropriate semantic conditions/environment. Due to its power of suggestion, its signified ('signified' = spirit, meaning as opposed to 'signifier' = letter, expression) often leads to other signifieds and this process can be extended endlessly, an ultimate, transcendental signified being disqualified. That is why and how meaning can be created endlessly. In *Ethics*, Aristotle calls this practice "plasis" to refer to the enticing and infinite process/pleasure of generating meaning or the fact of being involved in a creative activity excluding the possibility of ending it. Meaning therefore functions like a "*natura naturans", that is, something perpetually under construction, evolving, never coming to an end. Accordingly, the semantic field theory and the semiotic status of the symbol explain the notion of ontogenetic process assigned to the metaphorical synesthesia. Besides, since each word is surrounded by a field of meaning, each synesthetic metaphor can use its field of meaning and that of another or others to develop extra meanings by means of their lexical vicinity. Finally, the collusion of these semantic elements enhanced by their dynamic fields of meaning in addition to their holistic perception of reality, their underlying semes (crypto-semes) and their symbolic connotations contribute to generate an exceptional semantic power. Then, the deployment of the synesthetic metaphor reaches a point where its meaning is generated exponentially, constantly and motu proprio, that is, on its own, by the very virtue of the ontogenetic process. Consequently, it appears that synesthetic metaphors are very prolific in terms of meaning. They underscore a complex semiotic process out of which meaning is created on its own. It is as if language were diligently and endlessly generating meaning and, by the same token, recreating itself. That is the modus operandi of the metaphorical synesthesia. In light of the aforementioned considerations, we realize that one of the major differences between a classical metaphor and a synesthetic metaphor resides in the semantic prolificacy. The former is less prolific whereas the latter is highly prolific and sophisticated because of the ontogenetic process, which nurtures a broad spectrum of semantic and hermeneutic possibilities. The metaphorical synesthesia is so linguistically cogent that it aspires to Cratylism or the attempt to motivate language by annihilating or, at least, reducing the gap between the signifier and the signified (s) to achieve Absolute Poetry. Precisely, Cratylism is the peak of symbolism as a poetic, literary school or trend. All sincere symbolists aspire to achieve it.

Cratylism

The word 'Cratylism' derives from Cratylus, Plato's teacher, a philosopher of the Greek Antiquity who was preoccupied by the necessity to optimize language. He posited that linguistic signs are motivated. He believed that there are connections between the signifier and signified(s). Therefore, for him, signifiers are referential since they refer to reality. In fact, Cratylus's implicit desire was to promote the efficacy of language. Plato used him as a character in one his Dialogues titled "The Cratylus". In this dialogue, there are two characters discussing on the motivation of language: Cratylus and Hermogenes. Cratylus's contention is that the signifier is motivated, sounds have a meaning whereas Hermogenes disagrees with him. Over time, the word 'Cratylism' was coined from this discussion to highlight the aspiration of some poets, artists to create the ideal language, a medium so refined and powerful that it is endowed with the ability to bridge the gap between signifiers and signified(s). For the sake of materializing this ambition, a number of linguistic devices have been experimented: *ideophones, onomatopoeia, neologisms, catachresis, quasi-systematization of connotations to avoid linguistic prostitution, *phonic mimologism (imitative harmony: rhyme scheme, paronomasia, homophone, homometric lines/verses, vocalic harmony, consonantic harmony, alliterations, assonance, specific rhythm patterns: isotopic rhythms, paratactic patterns, etc . . .), and synesthesia. Consequently, synesthesia bears within itself the seeds of the ultimate aspiration of symbolism: to maximize the power of language, poetry, and arts in general: Cratylism.

Example: "I drank her sweet words" (*Les Fleurs du Mal*, Baudelaire)

In this instance we have a collusion of three sensory modalities: gustatory (drank), tactile (sweet) and auditory (words). They deploy three inter-sensory isotopies "(1—"drank" = isotopy of liquid referring to taste; 2—"sweet" = isotopy of touch referring to form/texture; and 3—"words" = isotopy of speech referring to hearing) but systematically harmonized by the psychological law of totality to express a unique case of euphoria that cannot be conveyed in ordinary language. The linguistic signs used in this synesthetic metaphor are specially designed to bridge the gap between the signifier and the signified(s). Additionally, the semantic distance between each inter-sensory isotopy, their syntactic and semantic interactions, their underlying connotations, create a substantial amount of poetic shocks. These shocks are exponentially enhanced by the esthetics of the surprise because in normal circumstances one is definitely not expected to "drink [. . .]words". Therefore, the poetic shocks become endowed with the power to conceive, nurture and coalesce three semantic levels: literal/figurative and, ontogenetic. It follows that this synesthesia is so sophisticated and cogently generated that it stands out as a semiotic and stylistic masterpiece. Consequently, it manages to reduce the arbitrariness of language and the signified in particular. Finally, it appears that what the speaker says and the ethereal feeling he experiences successfully and semantically concur. This instance is an interesting case of Cratylism that symbolism endeavors to achieve through synesthesia.

COMPARATIVE & CONTRASTIVE OVERVIEW BETWEEN THE SYNESTHETIC METAPHOR & CLASSICAL METAPHOR

SYNESTHETIC METAPHOR:

1—**Semantic shift by semantic indirection**
2—**Generation of isotopies**
3—**Idiolectic pertinence: creation of poetic shocks nurtured by the collage (esthetics of surprise)**
4—**Conjunction of 3 levels of meaning: literal/figurative, holistic, ontogenetic**

- => **outstanding refinement of the expression + meaning generated endlessly and motu proprio (on its own)**
- => **polymorphic and prolific aspect of meaning coming with:**

Five (5) sensory perceptions with reality viewed in three dimensions;

- Idiosyncratic quality: letters, numbers, colors having a gender or inducing a specific mood, a psychological state;
- Reality inherently associated with specific colors, shades, tints, etc...
- Adjunction of an extra-sense leading to spatial perception granting the possibility to quickly see/perceive hidden items

CLASSICAL METAPHOR:

1—Semantic shift by semantic indirection
2—Generation of isotopies
3—Classematic pertinence
4—Two (2) levels of meaning: literal + figurative => refinement of the expression but limited.

The main features of the metaphorical synesthesia and the metaphor in general come down to the very understanding of what the metaphor is: a semantic bridge and crypt.

METAPHORICAL SYNESTHESIA & METAPHOR IN GENERAL: A SEMANTIC BRIDGE/ CRYPT

As a semantic bridge/crypt, the metaphorical synesthesia (and the metaphor in general) connects two or more areas apparently different in nature. This crypt is purported to be created, intuitively or heuristically discovered by poets and deciphered by semioticians or linguists who thus become cryptanalysts. The word 'crypt' stems from the Greek adjective 'Kruptos' meaning 'hidden'. It designates an underground room or vault beneath a church and used as a burial place. Thus, a crypt is by nature and status a hidden place, which, accordingly, needs to be found out. Applied to semantics, a crypt refers to hidden meaning(s). As a semantic crypt, the metaphor links two or more **seemingly incompatible** areas in order to highlight their hidden and common semes (elements of meaning), and suggest their connections to the reader or cryptanalyst. As a semiotic device, the metaphor prompts experts to scrutinize two levels: the surface and the depth. At the superficial level the connections between the two or more areas do not seem to exist. However, they do at a deeper level. We can use a concept of theoretical physics called 'wormhole' to explain the metaphor because it particularly fits in its semantic framework and reflects, just like the metaphor, on an apparently very far distance between two areas. De facto, in his theory of general relativity, Einstein has proven that space-time is a flexible substance. It can be warped/curved by gravity and huge massive objects, which opens room for the possibility of creating a short cut in space and time to bridge the gap between two areas very far away from each other. By virtue of this postulate, if we travel from point A to point B separated by 30 billions of light years, it will take us centuries to cover this distance. However, we can create a wormhole, a short cut/tunnel by warping space and time, which will enable us to reach point B from point A in a very short period of time. Likewise, the metaphor is a semiotic wormhole, a semantic short cut, tunnel or bridge that connects two areas apparently unreachable because they seem to be very far from each other meaning-wise. It follows that the role of the cryptanalyst is to diligently dig the linguistic fabric of the semantic abyss to disclose the concealed connections and bring their areas closer with this semiotic wormhole. Consequently, the metaphor entails three underlying processes: semantic (meanings), semiotic (gradual unfolding of meaning and suggestion of specific directions of meaning: isotopies, by examining the signs through which meaning is built for the purpose of communication) and hermeneutic (decipherment of meanings => exegesis/reading).

SEMANTIC PROCESS

The semantic process breaks down two distinctive meanings: literal and figurative developed by the principle of semantic indirection whereby one says something but means something else. That leads to two specific isotopies and, possibly, a few sub-isotopies in case of high *idiolectic pertinence. The semantic indirection also illuminates the *paradigmatic axis of human language. De facto, linguistics shows that language functions within two axes: paradigmatic and syntagmatic. The former is vertical and deals with interchangeable elements. The latter is horizontal and focuses on elements that can be associated with others within a syntactic framework dominated by contiguity. Therefore, since semantic indirection shows something but factually points to something else, it entails the possibility for the metaphor to adhere to the paradigmatic axis. It follows that this axis suggests interchangeable meanings, a spectrum of semantic interpretations, and semantic power pertaining to the metaphor.

SEMIOTIC PROCESS

Isotopies created by semantic indirection generate layers of meaning, a semantic hierarchy, and a semic dynamism/flow, which suggests the possibility of gradually moving from one level (superficial) to another (deep) to show how communication effectively unfolds. Therefore, the cryptanalyst is responsible for:

- Using his/her linguistic arsenal to decrypt hidden semantic connections pertaining to two or more apparently incompatible areas;
- Showing how the meaning and semes are being constructed from one superficial area to a deeper level;
- On the basis of these pre-requisites, 'establishing' the semantic bridge/crypt between these two areas, structuring and assessing its linguistic and semiotic basis.

HERMENEUTIC PROCESS

It deals with the ability for the cryptanalyst to grasp the semantic bridge, decipher and exhume it. Thus, an exegetic work buttresses a heuristic work and, under their synergy, the bridge becomes explicit. However, in case of ***metaphor in absentia** the bridge is highly sophisticated and then much more difficult to decipher than in a ***metaphor in praesentia**. The **metaphor in absentia** refers to a special stylistic device in which certain elements of comparison just suggest the analogy between a comparing item and a non obvious compared item that operates as if it were **absent**. Hence the justification of the denomination "**metaphor in absentia**" that indicates an apparent absence of connection between the comparing and the compared items. On the contrary, the **metaphor in praesentia** excludes the possibility of involving a compared item operating as if it were absent.

Example of metaphorical synesthesia: "I drank her sweet words" (*Les Fleurs du Mal*, Baudelaire)

In this instance we have a collusion and collision of three modalities: gustatory (drank), tactile (sweet) and auditory (words). They are different in terms of meaning and do not seem to have affinities with each other. Besides, at a first sight, comparing what can be tasted (drank), touched (sweet) and what can be heard (words) seems irrelevant, or even nonsensical. However, if we pursue the analysis at a much deeper level, we realize that the gustatory, tactile and auditory modalities share common natural traits: all of them fall under the category of sensory perceptions. They have frequencies, electromagnetic properties, and coalesce to create a cross-modal metaphor. As sensory perceptions, they undergird cognitive elements because they provide us with information regarding the understanding and analysis of our reality, its peripheral environment, and human ecology. They deploy three inter-sensory isotopies:

1—"drank" = isotopy of liquid referring to taste;
2—"sweet" = isotopy of touch referring to form/texture;
3—"words" = isotopy of speech referring to hearing/audition.

Consequently, the metaphor decrypts the complexity of the human brain/mind set to accomplish specific tasks:

- Make subtle connections between apparently incompatible areas. As we mentioned previously, such connections are developed by a brain structure: the angular gyrus;
- Decipher hidden connections;
- Formalize/standardize them by means of a powerful language: a semantic bridge or crypt relating to two apparently distant and incompatible areas;
- Transcending apparent differences by highlighting, exhuming and synthesizing hidden similarities.
- Valorizing the eminently refined aspect of certain forms of language

It follows that probing and analyzing the complexity of the metaphor is:

- Analyzing that of the human brain/mind due to the correlations between thought/mind and language ((*psycholinguistics) => language is under the control of the left hemisphere of the brain);
- Studying how the brain functions to generate metaphors and poetry (*neuroesthetics: synergy of neuroscience and literature/arts);
- Examining the semiotic implications of the metaphor;
- As a corollary, examining the aptitude of poets (and artists) to create and use metaphors because they are eminently used by them (the metaphor being the most poetic figure. It is regarded as "the daily bread of poets" (Barthes)). Precisely, neuroesthetics seems to be the relevant tool to assess its complexities and come up with additional right answers.
- In light of these considerations, it appears that the human mind is an exceptional blending machine. It indeed associates ideas, concepts, notions, etc . . . However, there are two types of connections: simple and highly refined. The latter are hidden in the universe. Precisely, the brain of genii/artists seems to be predisposed to decrypt these very special connections or make them, which precisely endows them with the ability to create, decipher metaphors, or express themselves in a metaphorical language. Their angular gyrus, a small brain region, is highly conditioned to accomplish this remarkable task: setting up a semantic crypt /bridge between areas apparently incompatible by nature. Additionally, research has proven that:

    a—Creators', artists' brains have more connectivity because communication processed from neurons to neurons and bridged under the form of electrical signals via synapses all the way down to the thalamus (limbic system) is very intense. Such connectivity seems to endow artists and synesthetes with the aptitude of expeditiously solving problems, deciphering hidden connections, getting involved in creativity;

b—Synesthetes possess a kind of extra-sense, a sixth sense, which enables them to have a holistic perception of reality entailing the ability of perceiving items in three dimensions, through sight, audition, taste, smell, touch, gender, with vivid emotion and, more importantly, decrypting hidden objects (geometrical figures, etc . . . see **Ramachadran and Hubbard**)

These two factors can illuminate their ability to create/grasp metaphors and their latent complexities. As research grows in the field of neuroesthetics, neuroscience on synesthesia, we will certainly acquire more knowledge on the essence and modus operandi of the metaphor, which again surmises that a deeper understanding of the metaphor will unquestionably require the conjunction of neuroscience and literature. Consequently, to summarize, we can state that metaphorical synesthesia (and the metaphor in general) is a semantic bridge/crypt through which subtle connections are decrypted or created. It abides by the paradigmatic axis of language. Its analysis requires the synergy of neuroscience and poetry, and deploys a powerful and complex language reflecting the very complexity of human brain and mind.

e.g. This lady is a flower

The following is the illustration of the semiotic wormhole

SEMIOTIC WORMHOLE

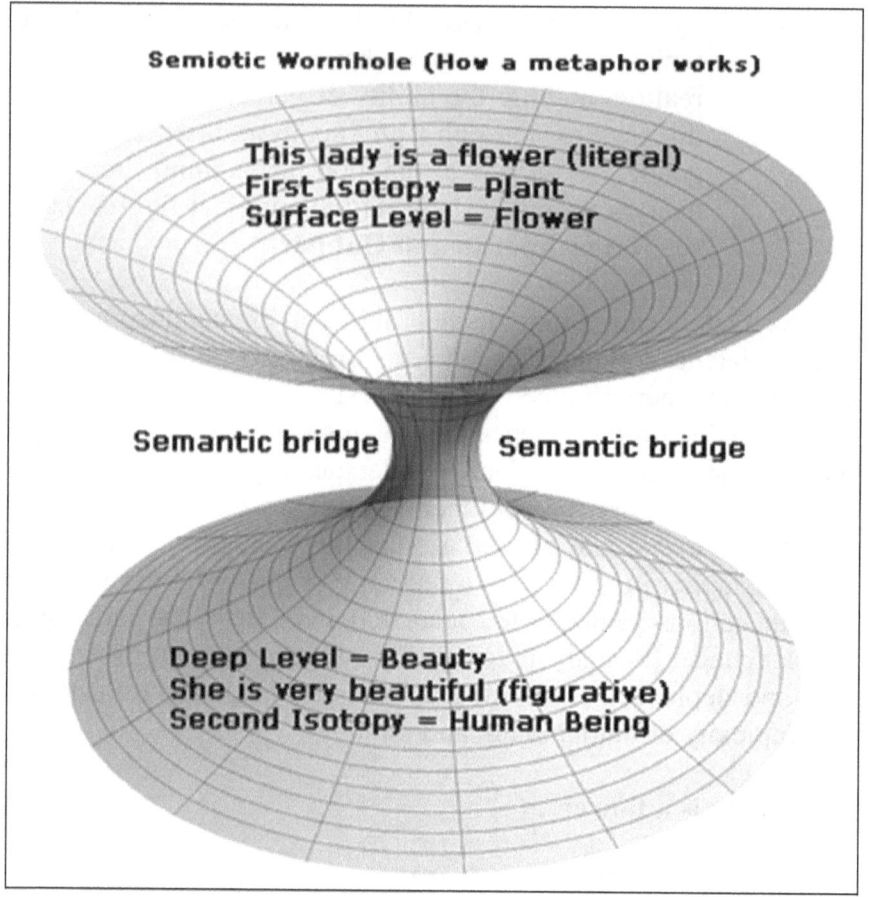

-> The metaphor transcends the surface and highlights the depth to reach the semantic nucleus or core of the meaning(s): the similarity of status: the lady = beautiful, the flower = beautiful

-> Here it is noteworthy that at a deep level, the lady and the flower connect through <u>beauty</u> since beauty is housed by both of them => the fabric of the semantic distance between them is thus <u>bridged</u>, <u>reduced</u> or even <u>warped</u>. Hence the concept of the metaphor viewed as a short-cut comparison or, better, a <u>semantic bridge</u>.

-> The passage from the first isotopy to the second creates an internal distension of the metaphorical discourse and, accordingly, enhances its semantic power and prolificness.

The analysis of the metaphorical synesthesia provides a metrical framework. From it, it becomes possible to measure synesthesia with specific canons. We coin it 'synesthesiametrics'. To establish synesthesiametrics, we have examined the list of the important canons of synesthesia and selected the most pertinent ones out of it. Therefore, those that best fit in the assessment of the metaphorical synesthesia can be used as its benchmarks. Here are they.

SYNESTHESIAMETRICS:

Synesthesiametrics can be assessed on the basis of 3 or 4 major esthetic canons: 1—the epistemological complexity of the metaphor, 2—its poetic quality, 3—semantic power, and 4—idiolectic pertinence.

This metrics can be re-written as follows:

SYNESTHESIAMETRICS = EPISTEMOLOGICAL COMPLEXITY x POETIC QUALITY x SEMANTIC POWER x IDIOLECTIC PERTINENCE

- **Epistemological complexity (E. CP.)** entails the <u>possibility of synergizing neuroscience and poetry</u>, which enables to show how synesthesia works by <u>means of neuroesthetics</u>. As previously mentioned, neuroesthetics is a field of knowledge bridging the gap between science and literature (or arts). It enables researchers to assess poetry and other arts in light of scientific notions (neuroscientific canons)
- **Poetic quality (P.Q.)** is governed by <u>the esthetics of the surprise</u> and the ability to generate <u>poetic shocks</u>. The metaphorical synesthesia is so innovative, original that it becomes regulated by <u>the esthetics of the surprise</u>. De facto, the collage, pre-requisite at the very basis of synesthesia, leads to the generation of poetic shocks, which enhances its poetic quality. Poetic shocks are <u>proportional to the semantic distance between "mimesis/referentiality" and "rhetoricity" as shown in this diagram. Accordingly, the greater the semantic distance is, the more poetic shocks can be generated</u>. "Mimesis" refers to the imitation of reality or

referentiality (reality). However, here by "mimesis", we mean what one says and by "rhetoricity" what one <u>factually means</u>. The shift from mimesis (what one says) to rhetoricity (what one factually means) provides room for the exercise of semantic indirection. It contributes to create poetic shocks and, as a result of this, the esthetics of the surprise (it was previously defined). Usually metaphors and tropes in general vouch for semantic indirection and poetic shocks. Accordingly, if a metaphor or trope occurs in a text, there will also be room for the exercise of semantic indirection.

- **Semantic power or density (SEM. D.)** involves the <u>**coalition of two or more sensory modalities with their underlying literal, figurative and ontogenetic meanings**</u>. **This combination is the concrete manifestation of synesthesia.**
- **Idiolectic pertinence (IDIOL. PERT.)** stems from the power of suggestion and latent symbolism. Idiolectic pertinenence can climax into Cratylism. The power of suggestion and latent symbolism include the <u>**meticulous choice of words or expressions imbued with a suggestive effect**</u> **(vivid metaphors, catachreses, synecdoches or any other trope, suggestive lyricism, music, uneven, loosen, and free verses, neologisms, systematization of connotation to safeguard linguistic catharsis, phonic mimologism and, its corollary, Cratylism)**

***Notes**:

It is not always possible to scientifically assess, measure synesthesia in a text. To systematically do so, it would be necessary to have the synesthete or poet synesthete within easy reach and scan his/her brain to realize its modus operandi, how the synesthetic process unfolds in him/her. Another option of subscribing to this canon is to have explicit or implicit indications of scientific facts buttressing the synesthesia in a text. For this reason, we use the word "possibility". It follows that we will often skip this first canon if its possibility does not occur and assess synesthesia on the basis of canon # 2 (poetic quality), canon # 3 (semantic power or density), and canon # 4 (idiolectic pertinence).

DIAGRAM OF POETIC SHOCKS

e.g. Her smile is singing

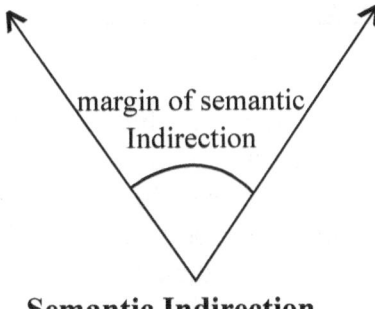

Her smile is singing (mimesis)   She is really gorgeous (Rhetoricity)

margin of semantic Indirection

Semantic Indirection

HERMENEUTIC CODE OF SYNESTHESIA:

Met. S -> Σ (at least 2 sensory modalities) + Prod (3 semantic levels: literal, figurative, and ontogenetic)

Metaphorical synesthesia can be re-written as:

| Met. S -> Σ (at least 2 sensory modalities) linked by a copula) + Prod (3 semantic levels: literal/figurative, holistic, and ontogenetic) |

The Combination of at least 2 sensory modalities + the Product of their 3 semantic levels

Where:

- "Met. S" stands for "metaphorical synesthesia"
- "->" stands for "can be re-written"
- "Σ" stands for "combination or summation"
- "Prod" stands for "product"
- "( )" stands for "of"

Given the fact that the metaphorical synesthesia is fundamentally a trope and that tropes have a number of variants, for the sake of explicitness, it is convenient to point them out and establish their classification and functions by taking their specificity into account.

CLASSIFICATION AND FUNCTIONS OF SYNESTHESIA VARIANTS

Synesthesia and their variants have simple and sophisticated constructions. It is noteworthy that most instances to illustrate the different functions of synesthesia in this section are inspired from Baudelaire, the precursor of symbolism and Rodenbach, an authority of Belgian symbolism. Indeed, Rodenbach's *Bruges-La-Morte* proves to be a highly refined symbolist work. It is a catalog of most of the finest synesthesia of symbolism. Mallarmé, the master of French symbolism, accredited this masterpiece and praised Rodenbach. Here is what he said: « *En lisant ses livres,* disait-il, *on a l'impression de la sensation fugitive, fixée, piquée, qu'enveloppe et cristallise la phrase sous une forme définitive* ».

So, most examples chosen from Baudelaire and Rodenbach are justified because they pertinently represent symbolism.

A—SIMPLE CONSTRUCTIONS

They are made up of metaphors in absentia, that is, metaphors in which certain elements of comparison just suggest the analogy between a comparing item and a non obvious compared item that operates as if it were absent.

Metaphors in absentia

e.g. 1 **Hughes did not hear this pain** (*Bruges-La-Morte*) -> the analogy between the comparing item: auditory perception (**did not hear**) and the non obvious compared item: tactile perception (**pain**) functions as if it were absent.

e.g. 2: **He drank that sound** (*Bruges-La-Morte*) -> the same remarks apply to this instance. Simple constructions are also composed of metaphorical synesthesia with adjectival phrases.

Metaphor in absentia with adjectival phrases

e.g. 1 A contagious silence (ibid.);
e.g. 2 a colorless silence (ibid.);
e.g. 3 a wounded voice (Depestre, *Etincelles et Gerbes de Sang*)

Simple constructions are theoretically simple. Actually, from the point of view of their meaning, they are complex. Based on the semantic complexity of their lexical items, we realize that they have a latent symbolism, especially those constructed by means of a metaphor with adjectival phrases. Thus, their morphological simplicity is contrasted by their semantic complexity.

B—SOPHISTICATED CONSTRUCTIONS

Sophisticated constructions can be found in catachrestic synesthesia, metonymical synesthesia, synecdochical synesthesia, polyptic or impressionistic synesthesia, hypallage synesthesia, synesthesia en "style artiste", and synesthesia in adynaton.

1—Metaphorical synesthesia.

We have substantially discussed metaphorical synesthesia. We will therefore study other variants of synesthesia. However, it is important to note that human language functions with two axes: ***paradigmatic and syntagmatic**. The metaphorical synesthesia is determined by a paradigmatic axis (vertical axis -> substitution) whereas the metonymic synesthesia presupposes a syntagmatic one (horizontal axis ->

association). Since the former depends on the paradigmatic axis (substitution), this provides the poet the leverage to operate shifts of meaning (substituting one meaning with another) by semantic indirection. The latter depends on the syntagmatic axis (association), which gives the poet/cryptanalyst the possibility to associate one meaning with another by contiguity through semantic indirection.

2—Catachrestic synesthesia: a—definition and modus operandi

Catachrestic synesthesia is a form of synesthesia whose internal structure is governed by a stylistic device called catachresis. A catachresis is a metaphor embedding a lexical figure, which means, molded by the coinage of a new word/new constructions borrowed from another nature or species to suggest a reality, idea, notion or concept quasi indescribable when using common language. Hence arises the linguistic necessity to create a new word to express something very special that ordinary language fails to describe by using specific *lexemes. Thus, by means of lexico-semantic audacities poets come up with a catachrestic synesthesia, constructed on the basis of a catachresis, a lexical figure.

e.g 1. a smoke of sounds (Rodenbach, *Bruges-La-Morte*),
e.g. 2   a voice of the same color (Rodenbach, *Bruges-La-Morte*)
e.g. 3   Colors that re-echo (Baudelaire, *Les Fleurs du Mal*)

b—Functions of the catachrestic synesthesia

The catachrestic synesthesia has an *apophantic function, that is, conceived to create by the "magical power" of art and poetic language. It enables us to see how effective language can be used to maximize communication. Accordingly, this function is primarily semiotic and poetic since poetry etymologically refers to creation, making (From the Greek *poiesis*—ποίησις: creation, making). In these two instances we have a catachrestic synesthesia. In light of the definition of catachresis, it dawns upon us that **"smoke of sounds** and **"voice of the same color"** are words coined poetically. They are employed to suit very special situations and notions that cannot otherwise be suggested by ordinary language. In "smoke of sounds" we have the coalition of two inter-sensory isotopies: sight (smoke) and hearing (sounds) to suggest ethereal realities. Likewise, "a voice of the same color" can be broken down into two inter-sensory isotopies: hearing (voice) and sight (color). Here the two isotopies are artfully synthesized to suggest the cryptic analogies existing between the material and the transcendent, the terrestrial and the supra-terrestrial and conveyed by the power of a catachrestic synesthesia to suggest a unique fact.

With respect to **"colors that re-echo"**, the visual mode **(color)** and the auditory **(re-echo)** coalesce to create an innovative form of color with special texture, a dazzling effect which can be suggested by nothing but a catachrestic synesthesia.

3—Metonymic/synecdochical synesthesia: a—definition, modus operandi

This synesthesia is embedded by a metonymy, a form of metaphor based on the concept of contiguity. "The word "metonymy" means "name change" in Greek (**Nomo: name and metein: to change, to associate**). *The Dictionary of Literary Terms and Literary Theory* glosses over it as follows: "In it (metonymy), the name of attribute or a thing is substituted for or associated with the thing itself. Common examples are 'The Stage' for the theatrical profession; 'The Crown' for the monarchy;". It is clear that since the stage (a part) is taken for the theatrical professions (the whole), metonymy takes a part of something for the whole, a piece for the set and, by the same token, indicates a process of contiguity. In this particular area, metonymy and synecdoche overlap because, precisely, synecdoche is a device in which the part represents the whole whereas the piece/ element represents the set, and thus, something else is underlying within the thing mentioned ('sun' in Greek = set, conjunction, together and '**dochein**' = to take up, to take over). Sometimes, the signified is taken for the signifier, the cause for the effect, and the abstract for the concrete, etc . . . Consequently, the difference between metonymy and synecdoche is minor or even negligible. However, synecdoche is likely to be less broad than metonymy. Given that the difference between these two tropes is almost negligible, we will just focus our attention on metonymy and consider that what applies to the former can similarly apply to the latter.

e.g. 1   **The radiance breaks out**
e.g. 2   **I hear the soft night that walks**

The first instance presents an inter-sensory metonymy composed of two modalities: 1—a visual sensory perception (radiance); 2—an auditory sensory perception (breaks out). The visual is a portion of the whole. Indeed the word 'radiance' designates a part (= part or more specifically the effect of the sun) whereas the sun designates the whole (cause). Accordingly, the part or the effect (radiance) stands for the whole or cause (the sun) in the sense that the sun is replaced by the radiance. In this respect, let us just bear in mind that metonymy etymologically means 'name change' -> **'sun' is <u>substituted, changed</u> with 'radiance'**

Likewise, instance # 2 includes the substitution of 'night', which stands for 'darkness' because it is changed with "night".

b—Function:

The metonymic and synecdochical synesthesia are variants of metaphorical synesthesia. So, they have the same functions as the metaphorical synesthesia, that is, optimizing a quality or emphasizing a deficiency by creating an internal distension of the discourse to generate a number of isotopies, which leads to semantic proliferation/power.

4—Hypallage synesthesia: a—definition and modus operandi

It is a form of synesthesia constructed out of a hypallage. The word "hupallage" derives from the Greek prefix "hupo": "underneath/below" and the verb "allassein": "to exchange". Therefore, the hypallage is a stylistic device in which one changes the word order in a sentence/line/text to establish a particular kinship between specific words when such a kinship should factually be established with others.

e.g.   "It seemed that one heard invisible wings, a parade of angels" (94, *Bruges-La-Morte*)

Here, the collusion of an auditory modality (**heard**) and visual modality (**invisible wings**) is highlighted. This sentence should normally be formulated this way: **"it seemed that one heard a parade of angels whose wings were invisible"**.

b—Function

The hypallage synesthesia is designed to operate a syntactic shift in the structure of a sentence to give it a different meaning by putting a special emphasis on the ideas suggested by the words that have been changed. Therefore, the syntactic shift brings about a semantic shift. As a matter of fact, the first sentence **"It seemed that one heard invisible wings, a parade of angels"** highlights a symbolism infused with several layers of meanings: hallucination, singular *concatenation of supernatural events, *hypnagogic experience, epiphany, ecstatic vision, etc . . . . On the contrary, the other sentence **"it seemed that one heard a parade of angels whose wings were invisible"** highlights less symbolism than the first. Consequently, the hypallage synesthesia usually provides special emphasis regarding an idea, concept, situation or notion and, accordingly, it has a stylistic function. However, since stylistics helps decipher meaning, the stylistic function becomes buttressed by a semantic function.

5—Polyptical or impressionistic synesthesia: a—definition and modus operandi

It is designed in such a way that it looks structurally and morphologically composed of internal paintings, landscapes or perspectives as if a master were drawing them with his brush. (From Greek: "poly" = "several" "opsis" = "vision, view, perspectives, picture, image"). It enables to express the gradual and quasi-imperceptible change taking place from one sensory perception to another, magisterially unfolding in little quasi-tangible touches, which gives the reader the impression that he/she is gazing at a painting, or an architectural masterpiece. Additionally, this synesthesia stresses the notion of view, perspectives, or personal impressions, light and color. It therefore relates to impressionism, an artistic school created at the end of the nineteenth-century in France. Impressionism highlights painting, light and color. What matters in it is not the ideas but impressions that things at which we gaze create on us. That is why the impressionist painter tries to suggest his impressions on the canvas/picture to communicate them to the reader/spectator. In the process the painter underscores these impressions through light and color. Here, the poet substitutes the painter to formulate his impressions through the use of synesthesia. *The Dictionary of Important Theories, Concepts, Beliefs, and Thinkers* provides an illuminating definition of impressionism in these terms:

> Movement in painting, especially in France, from the 1860s through 1880s, whose practitioners were concerned about creating spontaneous works observed from nature. Impressionism was never a unified artistic school with a coherent program, but rather a result of the common passions and ambitions of a group of young, aspiring artists[. . .] The early impressionists took their inspiration from Gustave Courbet's depictions of ordinary life and from the carefully observed landscapes of the Barbizon School. Their style was distinguished by fragmented brushstrokes and an intense palette of color, their objective to capture the immediate impression of the effect of light and movement in a scene [. . .] The movement's name was coined at the first independent shows, in 1874, derived from Claude Monet's painting *Impression—Sunrise* and from a hostile critic who dismissed the artists as "impressionists" who produced unfinished work that dwelt on subjective impressions. Other members of the group were Paul Cesanne, Edgar Degas, Camille Pissaro, Auguste Renoir, Edouard Manet. [. . .] Applying new theories about visual and physiological responses to line, color, and luminosity, Seurat and others developed *pointillism*, in which a picture built up from small dots of paint that resolve into recognizable forms when viewed as a whole, and divisionism, the separation of colors through strokes of pigments to create a scene of balance and statis. (195-196)

e.g. 1 A crystal voice that is singing, is broadening itself into halos, eddies where man yields, whirls, and gives in.

This synesthesia turns out to be a physical or psychological landscape structured into four (4) highly refined specific steps. Each of them comes up gradually: 1—a crystal voice that is singing -> auditory modality; 2—is broadening itself into halos -> visual modality;

3—eddies where man yields -> tactile modality; 4—whirls and gives in -> visual modality. Each step deploys a vision/perspective/picture and evokes fragmented brushstrokes as if the artist were trying to capture or underscore impressions that such a vision, enhanced by color and light, develops. Life seems to unfold out of pictures built up from dots and separated by colors of strokes. These impressions strikes the soul of the poet. Precisely, he endeavors to communicate them to the reader. **This particular aspect of impressionism** leads to *pointillism and *divisionism as it was conceived by Georges Seurat, one of the masters of this artistic school. In the long run, the superposition and synthesis of the four steps constitute a real impressionistic work of art.

e. g. 2 The music of snakes and ophicleides increased in pitch, carried along the frail wreath, in a sporadic tempo, with the song of the soprano.

1—The music of snakes and *ophicleides (obsolete bass instrument of music with keys) increased in pitch -> auditory sensory perception highlighting a supernatural scenery;

2—carried along the frail wreath -> tactile sensory perception highlighting channeled or organized violence;

3—in a sporadic tempo, with the song of soprano -> auditory sensory perception emphasizing the internal complexity of music. Each of these constructions suggests a delicate page of the soul, undergirds a range of supernatural thrills adjusted in crescendo. It is tactfully 'pregnant with' subtle lyrical modulations. In the final analysis, they translate an emotional landscape and the organization of the whole set contributes to mold an artistic perspective likely to touch the reader's psyche and convey the poet's impressions.

b—Function:

The function of the polyptical synesthesia is to transpose poetry into painting. It hones the symbiotic relationships that exist within the arts. As a matter of fact, most arts "speak to each other" (symbolist conception of art) because they are inter-connected in the universe and life in general. It appears that, the critic and any sincere truth-seeker, or scholar can, to a certain extent, use poetry or another art—as an hermeneutic framework—to explain painting, music, architecture, and other arts to decipher their hidden meanings. Likewise, a painter can use music to gloss over architecture. To reach such a goal the critic should empathize with the painter, musician, or architect. Here "empathize" has an etymological meaning, that is: "to internally feel together with, to suffer together with the artist". In other terms, the critic should provisionally "borrow the soul" of the artist to feel, to experience realities as if he/she were the artist stricto sensu. Consequently, polyptical synesthesia corroborates the fact that there are subtle correspondences between the arts and, precisely, synesthesia attests to, symbolizes these correspondences (interconnectedness). In the Salon de 1846, Baudelaire valorized this standpoint. He strongly believed that arts belong to a system of communication within which it becomes possible to use ones to explain and comprehend others. His *cosmogonical conception of creation as one Monolithic unity could have been applied to painting as well as poetry, music, or architecture. He stated: "The best critique is [. . .] what this painting will be, pondered by an intelligent and sensitive mind. So, the best report/comment on a painting can be a sonnet or an elegy." Arts are then assigned an exegetic or hermeneutic function because they are used as a foundation from which they explain other arts. This conception of using an art as an exegetic framework is also known as meta-discursive, a discourse/text commenting, elucidating another discourse and formalizing it. In light of these considerations, polyptical synesthesia shows the interconnectedness existing in arts. Therefore, its function is hermeneutic and artistic.

6—Adynaton synesthesia: a—definition and modus operandi

It is a synesthesia built on a device called 'adynaton' (the plural is adynata). Adynaton refers to a situation, an idea impossible to materialize because of its fictionism or idealism. It can also refer to a mystery, an enigma depending on the context. Anyway, the context is supposed to elucidate its meaning. 'Adynaton' derives from Greek and means 'not possible.'

> e.g. **music was trickling from pipes, drowning, erasing writings (Rodenbach, *Bruges-La-Morte*)**

Here three sensory modalities respectively auditory (music), tactile (trickling), and visual and tactile (erasing) coalesce to constitute an adynaton. It is elaborated by the situation described in the three inter-sensory modalities, factually belonging to the realms of the mystery, enigma. Indeed, in realistic circumstances, an isotopy based on liquids (trickling) clashing with one based on audition (music) pertains to the realms of the mystery because music deleting what is written remains a mystery. However, this phenomenon does not necessarily go against scientific laws because it may have a rational explanation but not yet found by scientific orthodoxy. It follows that the *semantic matrix of this synesthesia contains an algorithmic paradigm of the mystery or inexplicable. "Semantic matrix" also called "semantic nucleus" refers to the focal point from which the meaning of a word/words, a sentence, or a longer text originates. Regarding "algorithmic paradigm", it is a semiotic expression defined as: "a series of operations or events directed to a finality recognizable a posteriori in each operation or event." (Patillon, *Precis d'Analyse Litteraire,* 118). An algorithm can pertain to the area of the natural, mystery, technical, science, arts, etc . . . . Here, in each of these three operations: music trickling from pipes (1), music drowning writings (2), music erasing writings (3) there is a common denominator: the paradigm of the mystery. Each of them is single-handedly is directed to the mystery and recognizable as such. That is why this synesthesia is nurtured by an algorithm relating to the mystery. However, all adynata synaesthesiae are not based on algorithms, but, several tend to be. Besides, this particular synesthesia (as several adynaton synethesia) is enhanced by a kind of lexical drunkenness: an excessive use of words, special action verbs ('trickling', 'drowning', 'erasing') whose dramatic effect is to inject intense action into the synesthestic corpus and stress the concrete and physical manifestations of the mystery. The lexical drunkenness is not necessarily expressed by the quantity of words. It can be suggested by the semantic density or quality of words used as well.

b—Functions

The adynaton synesthesia has a uchronic/ideological function. The uchronic function relates to the way the poet/poetess conceives the world. He/she conceives it not as it is but as he/she would have wished it to be, according to his/her aspirations and ideals. The Greek prefix 'u' has a double meaning. It refers to something that is beside or outside, or just denoting absence or privation. It also designates the notion of the beautiful; and the noun 'chronos' means 'time'. The word 'Uchronic' therefore relates to anything located outside history and time, or beautiful/good. It follows that a uchronic vision is idealistic since it focuses not on what factually is but what is outside reality and history, space and time (because since Einstein it has been proven that both space and time are intimately linked: what applies to one ineluctably applies to the other) that is idyllic, utopic, and idealistic.

With respect to the ideological function of the adynaton synesthesia, it is usually—not systematically—employed in its pejorative connotation close to Marx's conception: "system of ideas formalized by a social group and purporting to accurately reflect reality but actually distorts it, creating a false consciousness that takes appearances for actuality." (193, *Dictionary of Important Theories, Concepts, Beliefs, and Thinkers*)

7—Synesthesia en style artiste: a—definition and modus operandi

It complies with the Decadent esthetics. The Decadent movement is the result of a schism that was created within the symbolist movement circa the end of the nineteenth-century. One of the most representative literary works of this movement is *A Rebours* by Huysmans. The Decadent movement values the factitious at the expense of the natural. It depicts the degeneracy of beings subject to the principles of heredity and dominated by forces of their environment. Decadent texts highlight preciosity. They have an eminently sophisticated style, steeped in the use of synesthesia, rare or far-fetched words. Sentences fraught with neologisms are a matter of course. They are usually composed of the adjunction of an affix (**in**égayées, **in**apitoyées) conveying the idea of suppression, ablation, declension, or fall; sentences, lines, verses or constructions in which the writer transforms an abstract force into a concrete force to make it act on beings. Moreover, one of the *iterative motifs of Decadent texts is ageing or senescence. Ageing or senescence strikes beings and acts on them in such a way that they are immersed into a state of passiveness marked with complacency and manifesting itself by means of two or three inter-sensory isotopies in which a passive verb or that which has a passive connotation is inserted; a personal pronoun ('lui') placed between the verb and the noun. Such is the idiolectic pertinence that features 'synesthesiae en style artiste'. Given their very high level of complexity and sophistication, this type of synesthetic construction is extremely difficult to translate from French into English. Thus, we will utilize a verbatim translation.

> e.g. 1 **Il sentit un brouillard contagieux lui entrer dans l'âme. He felt a contagious fog entering him into his soul.** = verbatim translation
>
> e. g. 2 **Une flamme lui chanta aux Oreilles A flame sings him in his ears.** = verbatim translation
>
> e. g. 3 **Un picotement lui vint aux yeux A prickling sensation came him into his eyes.** = verbatim translation

It is interesting to note that the verbatim translations suggest something appealing: a sizable amount of poetic piquancy inherent in 'synesthesia en style artiste', which enhances its idiolectic pertinence.

b—Function

On the basis of their high level of semantic refinement and stylistic sophistication (two or three inter-sensory isotopies into which a passive verb—or a verb having a passive connotation—is inserted; a personal pronoun ('lui') placed between the verb and the noun => 1—inter-sensory isotopy of touch (he felt) + 2—inter-sensory isotopy of sight (contagious fog) + the insertion of the personal pronoun 'him'+ the passive verb 'entering' + 3 another inter-sensory isotopy 'entering him in his soul') we consider that synesthesia en style artiste has artistic and apophantic functions. As a matter of fact, its idiolectic pertinence unquestionably suggests or maximizes the creation of masterpieces in their most refined forms. At the end of the general conclusion, it will behoove us to explore a few symbolist passages chosen among the most famous representative writers of this movement: Baudelaire, Rimbaud, Yeats, Valery, Mallarmé, and Rodenbach. This will enable us to illustrate symbolism with symbolist texts.

SUMMARY OF MAJOR LITERARY SYNESTHESIA VARIANTS:

1—The metaphorical synesthesia = embeds an analogy, a semantic bridge and is determined by a paradigmatic axis -> function: to transcend expression, reality and highly refine/ highlight it -> poetic/philosophical function
2—The catachrestic synesthesia = designed to form a lexical figure by coining a new word -> apophantic/lexicographic function
3—The metonymic /synecdochical synesthesia = determined by a syntagmatic axis -> function: to synthesize expression and highly refine /highlight it -> poetic/philosophical function
4—The hypallage synesthesia = based on the shift from one meaning to another -> semantic/stylistic function
5—The polyptic /impressionistic synesthesia = synthesizes several perspectives/views magisterially unfolding in little quasi-tangible touches -> hermeneutic /artistic/psychological function
6—The adynaton synesthesia -> uchronic/ideological function
7—The synesthesia en style artiste -> artistic/apophantic/ philosophical function

SUMMARY OF CHAPTER III

Synesthesia is a highly refined form of poetic expression resulting from a conjunction of different sensory modalities to constitute an organic unity by means of the psychological law of totality. Synesthesia focuses on how the left hemisphere of the brain generates poetry under the aegis of specific stimuli and brain areas. The complexity of synesthesia requires the symbiosis of several areas of human knowledge: poetry, semiotics, and cognitive sciences (especially neurology, anatomy and physiology of the human brain, and psycholinguistics). Accordingly, since it shows how the left hemisphere of the human brain generates poetry, it is necessary to understand the meaning of certain terms corresponding to brain regions and their impact on poetry production. Therefore, this chapter gives brief explanatory notes of the terms: limbic system, left hemisphere, neo-cortex, psychic areas of the brain, and synapses. Besides, semiotics is used as a powerful tool of investigation to analyze synesthesia because it deals with the systematic study of signs within poetry and any graphic system to show how they function internally and externally for the sake of communication. Precisely, synesthesia is a poetic sign. It has a signifier and a signified. Thus, it becomes a heuristic object for semiotics to decipher how poetic communication concretely unfolds within the symbolist system. Synesthesia belongs to intersensory tropes or transmodal metaphors. **Example of synesthesia:** "she **heard** (auditory modality) **a gorgeous color** (visual modality)". Hearing is compared to sight as if sight were endowed with the capability of being heard like sounds (underlying analogy). In this instance, there is a combination of hearing and sight. It follows that the semes of hearing (**heard**) and those of sight (**gorgeous color**) coalesce and generate a substantial amount of semantic power resulting from their respective sensory modalities. The analysis of synesthesia complies with the semiotic principle of semantic indirection and can be formulated by a specific hermeneutic code. There are a wide range of synesthesia. However, they can be simplified into five major types: constitutional synesthesia or idiopathic synesthesia, pseudo-synesthesia, metaphorical synesthesia and its variants, acquired synesthesia or pathological synesthesia, and color-grapheme synesthesia. All of them can be examined literarily or/and scientifically (PET, EEG, MRI, fMRI, etc . . .). They have several functions: artistic, apophantic, lexicographical, philosophical, cognitive, psychological, and uchronic.

CHAPTER IV

INTERESTS/USEFULNESS OF SYMBOLISM/ SYNESTHESIA: ITS EPISTEMOLOGICAL RUPTURE

The study conducted in this book has provided a semiotic framework whereby it enables a deeper understanding of symbolism, synesthesia, and the mechanism of inter-sensory metaphors (their production, use, analysis, and semiotic modus operandi); a holistic study illuminating correlations between the structures of the brain (left hemisphere, neo-cortex, temporal lobe, psychic areas of the brain, limbic system, synapses and neurons, etc . . .) and the generation of poetry/metaphors through the angular gyrus. This study proves that the brain can produce poetry under specific conditions. It also provides a minute assessment of the efficacy of inter-sensory metaphors as opposed to classical metaphors by means of a number of concepts and principles: idiolectic pertinence, poetic shocks, semantic indirection, to mention but a few. It explains techniques conceived and implemented by symbolism to maximize the efficacy of language and/or reduce the arbitrariness of the signified under the aegis of Cratylism. It opens the door to new forms of therapy for cerebral dysfunction on the basis of how the major areas of the human brain work—brain plasticity in particular (cerebral massotherapy, musicotherapy, phototherapy). Accordingly, it sparks several interests: linguistic/semiotic, philosophical, technological, cultural, artistic, scientific/cognitive, didactic, and therapeutic.

1—Linguistic/semiotic and teleological interest

Through symbolism synesthesia nurtures a highly refined literary expression, poetic by nature, and conceived to ward off or at least reduce the arbitrariness of language with a fruitful symbiosis between the signifier and the signified. Human language is limited in its ability to effectively express thought, emotions, feelings, and our vision of the universe. Synesthesia turns out to be an audacious and technical instrument conceived to get rid of these linguistic limitations. By the same token, it suggests the spectrum of technical possibilities into which poets, linguists, semioticians and all those who are interested in language can tap to maximize it. Such a posture culminates into Cratylism with its technical devices: phonic mimologism, imitative harmony, psychological rhythm (as opposed to metrical rhythm), idiolectic pertinence, onomatopoeia, musical silence, *proxemics, etc . . .

Cratylism entails a new conception of language designed to optimize the dynamics of thought. Since the mastery of the universe is linked to the power of language and thought, Cratylism definitely takes us a step forward to its mastery. It unquestionably endows language with special dignity, "ethereal wings", and a refined freedom. In fact, Cratylism can be so powerful that it gives the impression that language borrows from the divine.

Synesthesia is teleological in the sense that it has an end, a purpose. As a matter of fact, the sovereign and noble mission of human life on earth is to govern the universe by decrypting its mysteries so that we, humans, can enjoy existence in its fullness. Science is one of the means we can use to reach this goal. As a matter of fact, science reflects on laws presiding over phenomena by focusing on their 'how' in order to have a clear understanding of them. However, science, by itself, cannot finalize this cognitive enterprise. It needs an adjuvant, language and arts. Hence, the necessity of synesthesia (language, philosophy, and arts can fit in synesthesia because it synthesizes them) to buttress science. Indeed, if science focuses on the 'how', synesthesia (language/philosophy and arts) focuses on the 'why'. Language/art/philosophy constitute a triptych that science may use to refine its potential to govern the universe. By language/art/philosophy, we also mean creativity, imagination. Synesthesia is precisely enhanced by language, philosophy, art and creativity/imagination. Genuine, bona fide scientists have always valorized art and imagination. It is noteworthy that Einstein, for instance, prioritized creativity/imagination. In an interview he stated:

"Imagination is more important than knowledge." In that, he assigned it a very high level of importance. Most eminent scientists and thinkers of the past from Socrates, Plato, Aristotle, Descartes, Leibniz, Kant, Rousseau, Diderot to Swedenborg highly appraised imagination. They were artists, creative thinkers and scientists simultaneously. Let us briefly examine but a few cases to show that artists/philosophers used to be scientists as well. Plato, for instance, was both a philosopher and a mathematician but he was generally referred to as a philosopher. In the *Little Book of Mathematical Principles* Robert Salomon provides the following portrait of Plato: "He was also a mathematician and held the discipline of math in high regard. Above the gate of the Academy was written: no one ignorant of geometry can enter here . . . One story is that he invented a device with moving rods for the problem of "doubling the cube". On the other hand, his influence on the philosophy of mathematics was enormous." (32). Regarding Aristotle, he was a genius factually at home in almost every field: math, astronomy, botany, medical science, theater (he created the major laws of the Greek tragedy and most Greek dramatists: Sophocles, Euripides, Xenophon, etc . . . applied them) logics, metaphysics, and physics. With respect to Diderot, Voltaire nicknamed him 'le pantophile', that is: he who is the lover ('philos') of everything and every subject because he virtually knew everything, all ('pan' in Greek)).

Over time discrimination between science and arts occurred artificially. Such was not the case from the Antiquity to the seventeenth—eighteenth century. De facto, during the Roman/Greek Antiquity, there was no fine line between the various disciplines compounding human knowledge. Most disciplines nurtured symbiotic relationships with each other. For instance, the borderline between mathematics, philosophy, physics, even religion, to mention but a few, was nebulous enough. Let us just keep in mind these words of Plato: "Only he who is a geometer is allowed in this room." By "geometer", Plato neither meant a mathematician nor a specific expert of a given field because, formerly, knowledge was not divided into bits and pieces. On the contrary, it was initially a strong edifice made up of parts harmoniously connected with one another and the metaphor of the architectonics was used to designate such an edifice. Therefore, Plato was rather referring to anyone involved in the love of wisdom, knowledge, and critical knowledge. In other words, someone whose field of expertise covered a broad-spectrum: mathematics, physics, astronomy, astrology, cosmology, religion, philosophy, etc . . . However, later on, the underlying common edifice of many disciplines was dismantled and each discipline started functioning on its own. Subsequently, this brought about over-specialization, which, nowadays, has practically reached its acme. Nevertheless, even if we have reached overspecialization today, we should keep in mind that we need to think across disciplines to govern the universe, which amounts to seek truth through the synergy of the how (science) and the why (arts/language/philosophy -> synesthesia). Dr. Murray Gell-Mann, Nobel Prize laureate for physics, a great soul, supports this viewpoint. In his book *The Quark and the Jaguar* he cogently declares: "[. . .] Specialization, although a necessary feature of our civilization, needs to be supplemented by integration of thinking across disciplines." (12)

I strongly believe that art, in its most refined expression, is very complex and sophisticated. Its understanding requires a scientific approach or, at least, is conditioned by scientific principles.

Pythagoras believed that the universe was built on laws of symmetry, order, and principles of mathematics also called science of numbers. One of the most important of these principles was harmony also found in music, painting, architecture, etc . . . The founder of the universe was referred to as the "Great Architect". Therefore, for Pythagoras, the universe was a enormous work of art, a piece of architecture abiding by a pre-established harmony, designed and assessed by principles of geometry and specific numbers under the control of the Great Architect. Poets, musicians, painters were usually trying to take this harmony and science of numbers as a template or canonical and esthetic benchmarks. It follows that, arts did not subscribe to randomness but a highly organized and complex enterprise guided by scientific principles. Subsequently, artists systematically competed with architects by applying laws of harmony to their works. Such was the case of Ronsard, among several others. In fact, his 1550 edition of his *Odes* was structurally inspired by Pythagoras and his theory of cosmic harmony. Consequently, in light of these considerations, it dawns upon us that art is not the opposite of science but its dynamic and fruitful complementary. On the basis of this viewpoint, it can be inferred that synesthesia, which precisely embeds art, language and philosophy, is geared towards our enhancing the possibility of controlling the universe. Consequently, it unquestionably has a teleological interest.

Additionally, life without art is insipid and tasteless. Precisely, synesthesia is a bright manifestation of art in a wide range of forms. But for art, life would be equated with food without spice. Art wards off existential angst. It provides our life with substance and flavor. It is a patent fact of our potential as human beings to sublimate nature and access immortality. Last but not least, it enables us to peep into the splendors of the supra-reality and surreptitiously relish them. Through it, to use Baudelaire's formula, we catch a glimpse of these "gorgeous scenes concealed behind the grave." (Baudelaire, *New Notes on Edgar Poe* 1857) That was specifically how Proust and Baudelaire—among others—viewed art. So, synesthesia, as an art, definitely has a linguistic and teleological interest: conquer nature through its artistic, linguistic and semiotic power.

2—Philosophical and epistemological interest

Through synesthesia, we realize that all elements are interconnected or interwoven in the universe: all is in all. In fact, synesthesia successfully shows the connection and analogies between apparently unrelated domains and, in the process, provides a better and more substantial understanding of metaphors and the universe. By comparing domains or things that seem unrelated, synesthesia does prove that they are subtly related at a deeper level. Indeed, one of the basic laws of logics and science states that one can compare things or items only if they are of the same nature, for instance, a diamond and another or other diamonds. Thus, comparing things that are apparently unrelated proves that they are, indeed, related because they have implicit and cryptic similarities. For instance, sight, hearing, touch, taste, and smelling seem unrelated. However, sciences show that:

- All of them are sensory perceptions stemming from electrical signals bridged by synapses to neurons and neurons take these signals to the thalamus where they are subsequently transformed into specific information;
- They comply with specific laws of nature: they have frequency, wavelength, and period;
- They are endowed with an electromagnetic quality resulting from the interference of their internal vibrations and that of their physical environment;
- They can be measured/assessed scientifically by P.E.T., E.E.G, MRI, and other scanners. Consequently they have deep connections and, by the same token, they are similar, which makes them become comparable. What occurs in sensory perceptions can be extended to a cornucopia of domains in the universe. Let us take the case of man and a stone. they look different, and thus, not comparable. However, at a deeper level, they are similar
- They are all rigorously composed of the same building blocks of matter and the universe, that is: atoms whose structure is: a nucleus, protons with a positive charge, neutrons with no charge (protons and neutrons are made of 3 quarks), and electrons with a negative charge and revolving around the nucleus;
- They are all subject to the four major forces of the universe: gravity (to keep the internal components together, cohesively, preventing them from falling apart, and gravity carries "graviton", the hypothesized particle), electromagnetism (they are all surrounded by an electromagnetic field, which carries light -> photon), strong nuclear force (carries gluon), and weak nuclear force (carries IVB: Intermediate Vector Boson);
- They all subsume the building blocks of the universe and matter: atoms of hydrogen and carbon.

These building blocks come from the sun (a huge natural nuclear reactor constantly bombarded and torn by forces of gravity and those of fusion) because life and matter on earth originate from the sun. Consequently, man and stone look very different but, at a deeper level of analysis, it turns out that they are fundamentally similar in so far as they abide by the same laws of nature, basically have the same physical and chemical components, and stem from the sun. It follows that they are comparable and, therefore, analogous. We can extend these similarities farther and even in the realms of the human mind and concepts, we will come to the same conclusion. If we take reason and emotion for instance we can see affinities. For instance, medieval scholasticism used to oppose emotion and reason (abstract intelligence). As a matter of fact, reason/intelligence used to entail cognitive elements: such as memory, analysis, understanding, problem-solving, etc . . . In *The Birth of Tragedy out of Music* (*Die Geburt der Tragödie aus dem Geiste der Musik*) Nietzsche judiciously elaborates on these notions through two antithetical concepts: the Apollonian and the Dionysian. The apollonian epitomizes the rational, the logic and lucidly deliberate mind whereas the Dionysian stands for the irrational, emotional mind, driven by passion and emotion and never reason/intelligence. With Nietzsche these two types of conceptual models are crystal clear and respectively comply with reason and emotion according to classical canons. De facto, classical definitions of intelligence underscored their unequivocal differences. However, today, it appears that these definitions do not entirely function anymore because intelligence can also include non-cognitive aspects. Several influential researchers in the intelligence field of study had begun to recognize the importance of the non-cognitive aspects. For instance, as early as 1920 in *Intelligence and Its Uses*, (Harper's Magazine 140, 227-335) E.L. Thorndike used the term social intelligence to describe the skill of understanding and managing other people. Similarly, in 1940 David Wechsler described the influence of non-intellective factors on intelligent behavior, and further argued that our models of intelligences would not be complete until we could adequately describe these factors. Zahra Basseda et other researchers did likewise in *The Bar—On model of emotional-social intelligence* (ESI). Psicothema, 18, supl., 13-25.). On the basis of these considerations, we realize that there is no longer any watertight compartment between intelligence/reason and emotion. Both can share cognitive and non-cognitive elements. For instance, analysis, understanding, thinking, theoretically pertain to the cognitive field, not emotion. However, the intellective ability to understand, analyze, ponder the emotion (crying, sadness, anger, etc . . .) manifesting itself through the suffering of others involve emotion and reason as well. Consequently, emotion and reason/intelligence can be intertwined and thus, turn out to be a patent fact that the intra-cosmic unity impregnating many areas of knowledge and life in general can truly be enlarged further to encompass abstract ideas, concepts, human mind, etc . . .

It is noteworthy that what applies to sensory perceptions, stone and man, regarding interconnectedness, can also apply to a myriad of items and domains in the universe. Indeed, this ontological interconnectedness can be extended to an endless number of realms. Accordingly, most elements in the universe are interconnected and synesthesia precisely proves to be a powerful symbol of interconnectedness. As a symbol, it stands for the intra-cosmic unity in which its philosophical function resides. Actually, symbolism, as a philosophy and poetic school, attests to the fact that unity is concealed behind apparent multiplicity. The poet/philosopher is a seer, clairvoyant or magus entrusted with the mission of decrypting such unity and its underlying arcana. The philosopher/symbolist poet/fundamental mission is to decrypt hidden and subtle connections between the microcosm and the macrocosm and translate them into a poetic message. By means of this hermeneutic enterprise they have come to find out that the microcosm is a reflection of the macrocosm, the material mirrors the supra-terrestrial because as Hermes stated: "As above so below". Hermes's thought leads to the theories of correspondences at the core of which synesthesia finds its philosophical truth and function: decipher hidden analogies. With respect to the poet/poetess, he becomes a *cryptanalyst because he/she arrogates himself/herself the right to decipher universal analogies at the horizontal and vertical levels. In light of these considerations, it can be inferred that there is a typology of correspondences: horizontal and vertical:

- At the horizontal level: there are correspondences between elements of the material universe epitomized by synesthesia;
- At the vertical level: there are those between elements of the material universe and the supra-terrestrial. Those of the material reflect those of the spiritual. The best way to appreciate these correspondences is to study Platonism because the essence of this philosophy stems from the fact that the material world is nothing but an inferior, pale reflect of that of archetypes, which is sagaciously analyzed in Plato's *Myth of the Cave* (*The Seventh Book of the Republic*).

DIAGRAM FOR INTRA-COSMIC UNITY: (read in light of the analysis of the poem *Correspondences* by Baudelaire):

- HORIZONTAL LEVEL: SYNESTHESIA OR HORIZONTAL CORRESPONDENCES

"Perfumes, colors and sounds respond." (Baudelaire, "*Correspondences*")

Perfumes (olfactory modality) <=> **Colors** (visual modality) <=> **Sounds** (auditory modality)

S Y N E S T H E S I A

- VERTICAL LEVEL: VERTICAL CORRESPONDENCES: THE MATERIAL AND THE ETHEREAL/SUPRA-TERRESTRIAL

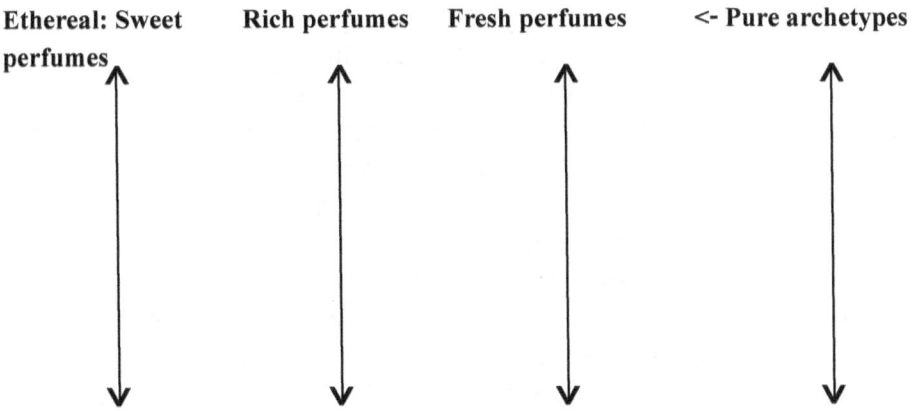

Ethereal: Sweet perfumes	Rich perfumes	Fresh perfumes	<- Pure archetypes
Material: Corrupt perfumes	Poor perfumes	Withered perfumes	<- Pale reflections

Besides, synesthesia operates a new epistemological break. It assesses a new status, theory, direction of knowledge, and a new vision of the world. Usually, we perceive reality in a divided, fragmented rather than a linear and unified way. With synesthesia, the perception of reality becomes de-fragmented and, accordingly, unified and holistic. This leads to philosophical reflections and questions: is reality shaped the way we usually perceive it (fragmented)? Or is it ontologically (that is: by itself) holistic, non-fragmented, linear or fragmented? Do we or do we not perceive it the way it is? Do synesthetes perceive it the way it really is? These phenomenological questions were adumbrated by Plato's Myth of the Cavern and need to be re-examined with the acquisition of knowledge on synesthesia, which gives way to a new epistemological perspective. So, synesthesia certainly brings us a new and fascinating world in so far as it enables, among other things, to compensate the fragmentation of reality and the universe. In light of all these considerations, my personal contention is that there is an ethereal and perfect consciousness that transcends the universe and oversees its pre-established harmony. We can call Him "God" or "Universal Intelligence" but His name does not matter, His function does. But for Him, there would not be a pre-established Harmony and reality would be very chaotic and inexplicable. It appears that the presence of this Ethereal Consciousness excludes any randomness because everything in the cosmos is subtly connected to everything. For instance, cosmologists had tried to find out the unified field theory on the basis of the analysis of the four major forces: electromagnetism, strong nuclear force, weak nuclear force, and gravity. This theory was supposed to provide a coherent and unquestionable understanding of everything both a macrocosmic and microcosmic levels. However, no one, even Einstein has managed to finalize such a Promethean enterprise. Only the first three forces have hitherto been analyzed under the "standard model" theory (gravity was excluded). It follows that this colossal endeavor cannot lie within the framework of four scientific forces, but within that of the complexity and the understanding of the Ethereal Consciousness overseeing the universe and its deep and outstanding unity. Precisely, synesthesia, with its cryptic and epistemological ramifications, turns out to be one of the cogent understandings and manifestations of this unity. It follows that synesthesia is purported to become one of the domains on which researchers of all fields should work in unison to tap into its range of possibilities, which will enable them to decrypt mysteries of the universe and render us, human beings, the lords of the cosmos. So, synesthesia has an epistemological and philosophical interest. It has a technological and scientific interest as well.

3—Technological and artistic interest

Synesthesia and symbolism have provided an interesting room for concepts and ideas to inspire modern technology through a number of technical inventions: the wheel, cinema, and video. All these inventions are based on the principle of the collage that was inherently connected to the key-concepts of symbolism and synesthesia. As we have mentioned earlier, the collage is the collision of distant realities, notions, ideas, concepts or items exercised by the magic of art and creativity to generate a powerful image. The farther these realities are, the more powerful the image becomes. The collage is an invention of symbolism but borrowed by surrealism. It has 'ignited' several technical inventions.

a—The invention of the modern wheel and the car

Guillaume Apollinaire, a French poet, dramatist and precursor of surrealism inspired by symbolism, explained the process that had led to the invention of the modern wheel in the preface of his play *Mamelles de Tiresias* (*Breasts of Tiresias*). This process was triggered by the symbolist collage through the collusion/collision of three seminal ideas: walking, a machine imitating walking, and an instrument enabling this machine "to walk" as if it had a leg. So, in his preface not only did he explain the mechanism that inspired the invention of the wheel but also coined the word "surrealism" that was going to be adopted later in literature and other arts. He declared: "Quand l'homme a voulu imiter la marche, il a créé la roue qui ne ressemble pas à une jambe. Il a fait ainsi du surréalisme sans le savoir." (*Mamelles de Tiresias*), which means: "When man wanted to imitate walking, he made a machine that would walk. Then, he created the wheel, which does not resemble a leg. So doing, he became a surrealist without knowing it."

b—The invention of cinema/3D movies and 3 D television sets

Cinema is fundamentally based on the principle of the collage. In it, four concepts are combined: motion (From Greek 'Kine': motion, movement), view, sound, and the opportunity of viewing images in three dimensions: length, width, height. From the inspiration of the symbolist collage, all these concepts were put together and translated into the creation of cinema, three dimension movies and television sets for the mirth of viewers and movie lovers. Synesthesia through symbolism has also inspired impressionism, cubism and, by the same token, helped in understanding and discovering the fourth dimension, that is time.

Consequently, it was at the forefront of scientific discovery. We will examine this aspect in the section devoted to the scientific interests of synesthesia.

c—Cubism, impressionism

Cubism values the techniques of assemblage (also known as the collage), viewing an object at every single angle, visual perceptions, to mention but a few. These techniques actually originated from symbolism, especially impressionistic/polyptical synesthesia that can be considered one of the arts that sowed the seeds of cubism and impressionism. The *Dictionary of Important Theories, Concepts, Beliefs, and Thinkers* confirms this perspective in its definition of cubism. Here is how it insightfully defines it:

> Centered mainly in France . . . cubism was pioneered by Pablo Picaso and Georges Braque and named by hostile critic, Louis Vauxcelles, after Henri Natisse's description of Braque's images as "cubes."

> Cubism was a stylistic reaction to the lush sensuality of impressionism and a visual response to the hard-edged mechanism of the industrial world. Its approach was inspired by the bold use of color in the late paintings of Paul Cezanne and by the geometric abstractions of African sculpture. It was founded on the idea that space is visually and conceptually ambiguous and that reality resides in the mind's perception of it. The cubist solution to the eternal problem in painting, that of depicting volume on a flat plane, was to see an object simultaneously from all sides, Picasso and Braque also introduced the use of collage in painting, adding materials such as paper, sand, wood, and cloth to a canvas, both to evoke the real world and to undermine the idea movement and set out its theoretical principles in his 1913 essay The Cubist Painters. (84-85)

In the same vein, Christopher Gray underscored the influence of symbolism on Cubism in his study *Cubist Aesthetic Theories*. He stated:

Relationships between cubism and those late nineteenth-century artistic movements generally called symbolist have been pointed out before, but usually on the level of aesthetics. Thus, for example *Professor Gray, examining the theoretical writings of Gleizes and Metzinger, the criticism of Salmon and Apollinaire, has been able to demonstrate that the essentially idealistic outlook of cubism, the belief that pure form, true reality exist beneath shifted appearances, is clearly allied to Symbolists's **search for preexistent or an ideal reality in the structure of the arts.** Of course the theory of correspondences developed by Baudelaire, practiced by Neo-impressionists and refined by Mallarmé and his followers, is of great importance to Cubists and, in turn, its indebtedness to the German philosophers has been justifiably emphasized (Kant, Schopenhauer). There are, however, other connections linking the generations associated with cubism and symbolism, connections that are primarily historical rather than aesthetic. Personalities and ideas of one group were known and understood by the other, fruitful contacts were made in a situation that was not only personal and direct but which was also governed by a fundamentally social program. (37).

d—Neuroesthetic and cultural interest

Synesthesia/symbolism has laid the foundations of human culture. Indeed, the meticulous and dynamic expression of the five sensory perceptions and their underlying connections show that each sense valorizes a particular aspect of our culture. We use sight to gaze at works of art: painting, architectural achievements, photographs, enthralling stars in the sky at night, etc . . . Hearing enables us to relish music. Touch grants us the opportunity to appreciate texture and forms of matter. Through smell we enjoy odors and perfumes. Regarding taste, it gratifies us with the treat of a wide range of culinary delicacies. The extra sense/ "sixth sense" inherent in synesthesia helps in the blending of these five senses with colors, numbers, size, height, gender, mood (since specific letters, numbers and colors "express" a specific mood, it is possible to talk of their idiosyncrasy) and decipher the mysteries of the unknown. Therefore, synesthesia eminently enhances the greatness of human culture and the unique opportunity to appreciate life holistically. It is expressed and manifested in all forms of art: poetry, painting, architecture, music, cinema, etc . . . Since all forms of art, when effectively conceived, are so powerful that they communicate us feelings/emotions: suffering, mirth, sadness, anger; a specific vision of the world; and even an esthetic framework, synesthesia posits a neuroesthetic and psychological interest. As Dr. Hugo pertinently outlines in his article *Art and Synesthesia*:

> Works of art are literally pregnant with meaning. The highest form of symbiosis between synesthesia and metaphor happens in art, because synergy is the essence of the living present and the essence of art. Basically, science examines and explains *'how'* and art provides a vision of *'why'*. Art points a direction, and science provides the transportation to get you there. Today, there is an increasing attention for the subject of synesthesia; online, in the media, in publications, conferences, symposia, scientific research, and in the arts. Synesthetic experiences are becoming more and more part of a daily awareness. Today, the ultimate synesthetic art form is still cinema. About the senses, Laura Marks wrote: in film, vision (or haptic visuality) can be tactile, *"as if touching a film with one's eyes"*, and further, *"the eyes themselves function like organs of touch"* (8).

> Through the art and technology of synesthetic cinema (expanded cinema) a filmmaker can, *"express a total phenomenon—his own consciousness"* (9). Synesthetic cinema, more than any other medium, has demonstrated a trend towards the *'polymorphous' (having many forms or functions)*—the intercorrelation of the four modes of human consciousness: thought, intuition, emotion, and sensation. Powerful combinations of computer multimedia, virtual reality, and holographic cinema hold the promise for further complete expression of the synesthetic experience. Is cinema becoming an individual form of art? At the moment, digital films are produced anywhere by anyone and are accessible

on the global network, as real-time experience, as interactive space, as interconnectivity. In the near future, our senses will become *'travelling senses'* and tele-synesthetic (10). New research indicates that neurological discoveries about brain function and responses to stimuli can offer insight into how the brain develops metaphors and interprets art. Neuroscientist Dr. Ramachandran emphasized that these theories and observations are *"no more than hesitant first steps toward a science of art—towards discovering artistic universals—the new science of neuroesthetics"* (11). (Hugo, Art and Synesthesia) Dr. Hugo summarizes his thoughts in this diagram:

Two ways to challenge the classic view of perception: art and synesthesia

Art is sensuous knowledge

Art and synesthesia are both the result of the united senses of the mind

The arts offer multisensory forms of knowing and communicating

A synesthetical approach to reality is one of the primal sources of art

In art one dimension is often evoked by another

Art makes new connections between the senses

Synesthesia appears in all forms of art

Synesthesia: the united senses of the mind

> *All art constantly aspires towards the condition of music.—Walter Pater*

Hugo pursues with his reflection:

> Music (as a form of expression) is a perfect symbiotic model of unity; matter, form and content are one. Music allows the individual to experience deep and intense emotions. Sound is a 'unifying' sense—creating an immediacy of effect. To some degree, all experience is synesthetic because the 'synesthetic experience' is the result of 'the united senses of the mind'. The synesthetic process may be clarified by the diagram below. The dotted lines starting at the terminus of each sensation (at the red arrowheads) and intersecting the bands or channels of various senses might be taken to represent the mechanism (scanning and combining) of the common sense. This diagram may be read downward and is self-explanatory. (Ibid.)

The following diagram gives an idea of the synesthesia process. It is based on Gino Caserne (2004) and was modified by Dr. Hugo Heyrman (2005). I slightly modified it as well (see the second diagram).

THE SYNESTHETIC PROCESS by Hugo and Caserne

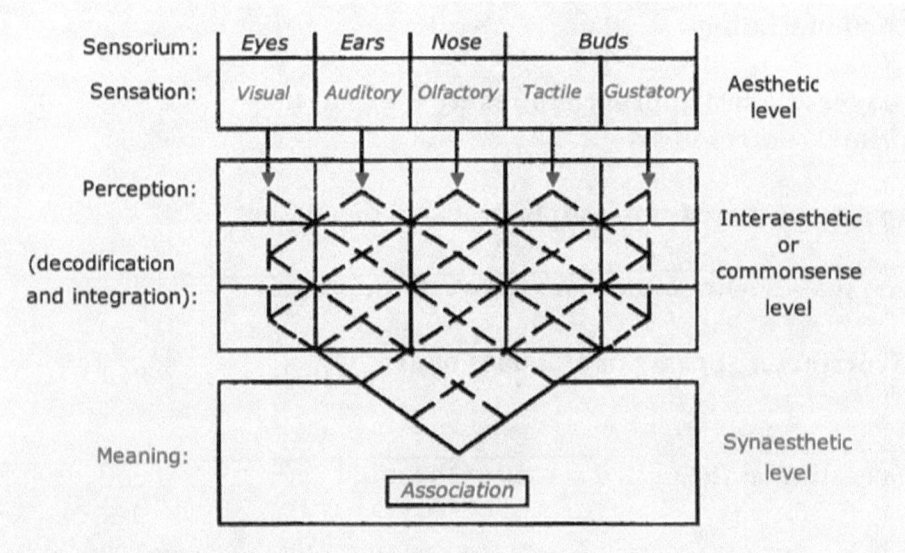

THE SYNESTHETIC PROCESS completed by Bonaventure Balla

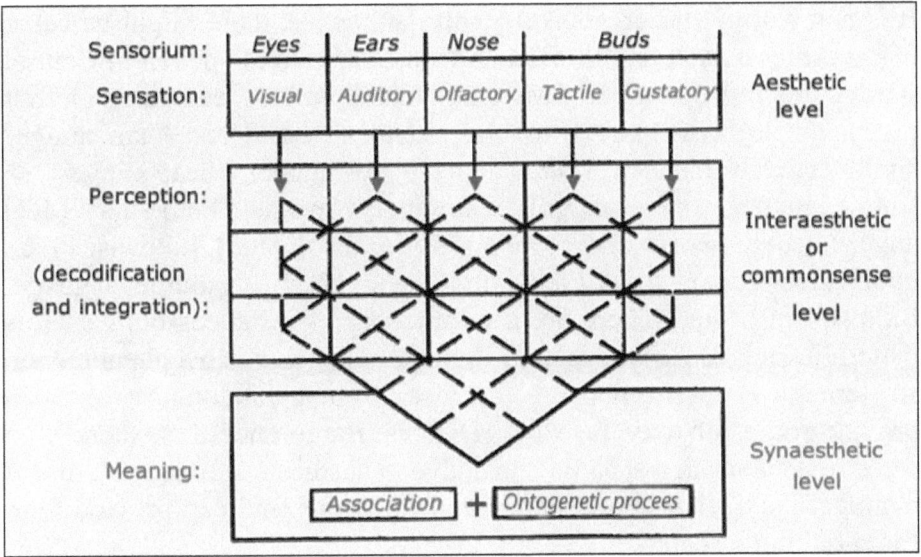

The diagram is incomplete because the box referring to the meaning at the bottom needs additional information -> at the synesthetic level, the meaning is governed by two factors: 1—association and 2—ontogenetic process. The ontogenetic process is missing here. It stems from the fact that synesthesia is endowed with implicit symbolism nurtured by lexical vicinity, the combination of sensory modalities, and their idiosyncratic manifestations presented in three dimensions. All these considerations create a meaning that is so powerful that it endlessly generates itself => ontogenetic process.

Association + ontogenetic process

The meaning is literal, figurative and ontogenetic: it is nurtured by lexical vicinity and consecutive latent symbolism => accordingly, it becomes so powerful that it endlessly generates itself, *motu proprio, like a *natura naturans (an open-ended nature)

Case of Bantu languages:

It is noteworthy that in several Bantu languages, there might be cases of synesthesia. Actually, in Beti and Basa languages spoken in Central Africa, the verb "hear" is always associated with at least three or four meanings: to hear, to feel, to understand, to be. Thus: I am hungry simultaneously means: "I am hungry/I feel hunger/I hear hunger" -> source language:[m'awok zje]. "I am angry" means: "I am angry/I feel anger/I hear anger" -> source language: [m'awɔk olun]. Likewise "I am good" means "I am good/ I feel good/I hear good" " -> source language: [m'awɔk m'b əŋg] There are a cornucopia of instances built on this pattern in Beti, Basa and several Bantu languages. Such a phenomenon so iterative is far from being the fact of sheer randomness and can provide room for deep research. Did our Bantu ancestors, those who were first to speak our languages have synesthesia ? If so, how did it manifest itself? These questions need to be dug and can provide food for enriching research.

However, not only did synesthesia/symbolism lay the foundations of human culture in general but also brought a significant contribution to modern literature and culture. Indeed, its major poets Baudelaire, Mallarmé, Rodenbach, Valery, Yeats, T.S. Eliot, etc . . . inspired modern literature and culture. So, it has a historical and literary interest.

4—Historical and literary interest: symbolism as a layer of the foundations of modern literature

Symbolism sowed the seeds of modern literature through several areas: the introduction of free, freed/loosen verses and free rhythm (that is: a rhythm conditioned by the psychological disposition of the author rather than the canonical disposition of prosody); concept of interconnectedness extended to all arts; transposition of painting into poetry and culminating into calligrams/hypertext; intertextuality; "death of the author"; creation of a new semantics on the basis of a new syntax (hyperbaton, inversion, rejection of narration); techniques of the mise en abyme (also valued by the "nouveau roman") also known as "internal reduplication" but enhanced by André Gide; and the valorization of music.

Introduction of free, freed verses and rhythm

Free and loosen verses were introduced by Baudelaire, Verlaine, Rimbaud, Mallarmé and Valéry. They constituted one of the main elements conceived to free poetry from classical and rigid esthetic canons, to catalyze creativity and innovation. With respect to rhythm, it became also freed under the influence of romantics (see Victor Hugo "j'ai disloqué ce grand niais d'alexandrin see chapter 1) and symbolists. In fact, rhythm was no longer subordinated to classical and rigid rules established and standardized by Boileau's *Art Poétique*. Instead, rhythm was now subject to the discretionary power of the poet/poetess. Therefore, it became internal, that is, contingent upon the psychological status of the artist, his mood and soul, not external and standardized rules. As a result of this, henceforth, symbolists conceived and adopted a psychological rhythm as opposed to metrical or canonical rhythm. The introduction of free, freed verses and rhythm and its unique exploitation by symbolists gradually paved the way to modern poetry and literature in general. Since free, loosen verses and rhythm were extensively discussed in Chapter 1, we will not conduct an unnecessary and additional examination of them here and this takes us to the interconnectedness of arts.

Interconnectedness of arts, transposition of painting/drawing into poetry

Symbolism equally influenced modern literature by illuminating the interconnectedness existing in arts especially painting, music, and drawing. This interconnectedness is a direct sequel, the very corollary of intra-cosmic unity (symbolized by synesthesia) since the macro-cosmos is the template of the micro-cosmos. In light of intra-cosmic unity, symbolist poets visually expressed certain themes of their poetry through a technique called lyrical *calligrams in which they merge or at least unified poetry and painting. Lyrical calligrams were a *pictographic technique that appeared for the first time in Mallarmé's famous poem titled "*Un Coup de dé jamais n'abolira le hasard*" ("A Throw of the dice will never abolish chance") and epitomized a remarkable osmosis between the signifier and signified, the form and the substance. The theme of the text was spatially and pictographically represented by a special array of words and lines of the passage. Thus, a poem became a privileged space where both the canvas and a graphic system were unified harmoniously and artistically. Calligrams were subsequently systematized by Guillaume Apollinaire in the twentieth century. Additionally, poetic texts were frequently exploited for their musical quality. Such was the case of Mallarmé's and Rodenbach's texts among others, used on account of their outstanding musical quality. De facto, Mallarmé's poetry has been the inspiration for several musical pieces, notably Claude Debussy's *Prélude à l'après-midi d'un faune* (1894), a free interpretation of Mallarmé's poem *L'après-midi d'un faune* (1876), which creates powerful impressions by the use of striking but isolated phrases. Maurice Ravel set Mallarmé's poetry to music in *Trois poèmes de Stéphane Mallarmé* (1913). Other composers chose to use his poetry in song include Darius Milhaud (*Chansons bas de Stéphane Mallarmé*, 1917) and Pierre Boulez (*Pli selon pli*, 1957-62). In the same vein, Rodenbach's *Bruges-la-Morte* was used by the composer Erich Wolfgang Korngold as the basis for his opera *Die Tote Stadt*. Consequently, There was no more watertight compartment between poetry and prose (the abolishment of the borderline between them was almost formalized by Baudelaire in *Petits Poèmes en Prose)*, music and poetry, painting and poetry. Mallarmé's earlier work owes a great deal to the style of Charles Baudelaire. His later *fin de siècle* style, on the other hand, anticipates many of the fusions between poetry and the other arts that were to blossom in the next century. Most of this later work explored the relationship between content and form, between the text and the meticulous organization of words and spaces on the page. This culminates into his famous poem, *Un coup de dés jamais n'abolira le hasard* of 1897. In an article published in New YorkTimes and titled "*Art Review; A Poet, Wit and Critic Heeded by the French*", Grace Glueck validates Mallarmé's outstanding contribution to poetry and modern literature. Her reflections are particularly revealing. She declares:

> Although his writing was more than occasionally proofread, Stephane Mallarmé (1842-98) was one of the most influential cultural figures of 19th-century France. An avant-gardist, a wit, a salon keeper, a fashion commentator, a translator of Poe's poems, a critic who supported the Impressionists and a forerunner of the Symbolist poets whose innovative ways with language had a significant impact on Modernism, he himself

was nothing less than a work of art. His closest friends were painters: Edouard Manet, Berthe Morisot, Pierre-Auguste Renoir, Edgar Degas and James Whistler. He defended their work in his critical writings, and they did portraits of him and illustrations for his books. Paul Gauguin, Henri Matisse, Edvard Munch, Odilon Redon, Auguste Rodin and Felix Vallotton, among others, also produced Mallarméan images and tributes. More or less pegged to the centenary of Mallarmé's death, an intense but unfortunately short-lived little show, "A Painter's Poet: Stephane Mallarmé and His Impressionist Circle," can still be seen (better hurry!) at the Bertha and Karl Leubsdorf Art Gallery of Hunter College. The show was put together by Jane Mayo Roos, associate professor of art history at Hunter, and a team of graduate students in conjunction with the Bibliotheque Litteraire Jacques Doucet in Paris. Its informative catalogue, with many scholarly contributions, is an essential part of the show. (6-7, *Art Review; A Poet, Wit and Critic Heeded by the French*, New York Times, March 12, 1999)

All this fruitful symbiosis between arts climaxed into the conception of theories and notions such as: hypertext/intertextuality, and death the author.

Hypertext/intertextuality, death of the author,

The theories and concepts regarding the hypertext, intertextuality, and death of author are frequently featured in modern literature. They were initiated by symbolists and Mallarmé in particular. In fact, Mallarmé was so ambitious in his poetry that he aspired to create what might be called a purely reflexive work, that is, a work so powerful that it arrogates itself the right of speaking by itself and, in the process, erasing the author by artistically killing him. Such a work—at least theoretically—no longer needs the author since it becomes self explanatory and self glossary. This theory had had a tremendous impact on the twentieth-century theoretical and critical apparatus and was commented by a number of authoritarian critics and translators. One of them, Barbara Johnson, has written an article entitled: *"Mallarmé gets a life"* in which she relevantly explained the theory of the poet's self erasure as it was proponed by Mallarmé. She said:

There is a reason for this erasure. The eclipse of the author by the work is not an accident of Mallarmé's criticism: it is Mallarmé's principal literary discovery. It was Mallarmé himself who dreamed of 'a text speaking of and by itself, without the voice of an author'. The affirmative erasure of the poet from the work was a goal for which he never stopped striving: 'The pure work implies the elocutionary disappearance of the poet, who leaves the initiative to words.' And it was Mallarmé himself who created the myth of his lack of biography: writing to Verlaine in 1885 in response to a request for a headnote for his poems, he spoke of his 'life devoid of anecdote'. Twenty years earlier, Mallarmé had announced to his friend Henri Cazalis, 'I am perfectly dead . . . I am now impersonal and no longer the Stéphane you have known, but an aptitude the spiritual universe has to see and develop itself through what was once me.' [. . .]

It was largely by learning the lesson of Mallarmé that critics like Roland Barthes came to speak of 'the death of the author' in the making of literature. Rather than seeing the text as the emanation of an individual author's intentions (always a probabilistic and speculative enterprise), structuralists and deconstructors followed the paths and patterns of the signifier, paying new attention to syntax, spacing, intertextuality, sound, semantics, etymology, even individual letters. The theoretical styles of Jacques Derrida, Julia Kristeva, and especially Jacques Lacan also owe a great deal to Mallarmé's 'critical poem.' Moreover, it has been established that much of Mallarmé's work influenced the conception of hypertext, with his purposeful use of blank space and careful placement of words on the page, allowing multiple non-linear readings of the text. This becomes very apparent in his work *Un coup de dés*. In each case, Mallarmé had been there before them: calling himself a 'syntaxer' and syntax the 'pivot of intelligibility', writing a book about the meanings of sounds and letters in English words, creating a concrete poem out of typography and position on the page, inventing a style of critical prose as well as poetry in which ellipses, discontinuities and obscurities played an integral part, and criticizing romantic subjectivity and bourgeois realism. Freed from conventions of coherence, authority and psychology, texts could be allowed to unfold as infinite signifying systems. This is not to say that Mallarmé's late, most stylistically radical texts have nothing to do with the desire for coherence. Indeed, one of the paradoxes of Mallarmé is that, along with his fragmentation of all the usual modes of meaning, he also imagined that 'The Book' would put everything back together in a higher synthesis. (12-14, *Mallarmé gets a life*)

The technique of mise en abyme known as "internal reduplication" is also part of symbolists' contribution to modern literature.

The mise en abyme or internal reduplication

The mise en abyme or internal reduplication was enhanced by André Gide's novels and especially *Les Faux-Monnayeurs*. It already existed before and was used by a number of authors such as Shakespeare in *Hamlet* (scene regarding the mousetrap) and in the Middle-Ages with Marie de France in *Lais* de Marie France and especially the lai called *Eliduc* where the mise en abyme is performed by an animal character, a weasel (p. 61). The weasel symbolizes immortality/ the transcendence of life upon death. As such, it is instrumental in transforming despair into hope and regeneration. In *Eliduc* the weasel metaphorically represents a woman (Eliduc's beloved) and inspires Eliduc, the protagonist. It reenacts the drama unfolding in the lai and helps decrypt the enigma with which Eliduc copes and this enables him to save his beloved from death.

However, the mise en abyme was highlighted by Belgian symbolists and particularly Maeterlinck and Rodenbach. The latter took it a step further. Indeed, in *Bruges-La-Morte,* one can find plenty of instances of the mise en abyme. Because of its recurrent pattern, it factually becomes standardized. The fact of the matter is that, in this poetic prose novel, the presence of mirrors stylistically formalizes the mise en abyme. Actually, the mise en abyme is also called "specularity" stemming from the Latin word "speculum": "mirror". The key-concept of this technique is to reproduce or reconstruct the plot (the way a mirror reproduces the image of objects). It consists in reproducing the plot at a microscopic level in such a way that a novel subsumes a "micro-novel", a play a "micro-play" and a poem a "micro-poem". In the final analysis, by exploring the micro-element of the literary work one can easily recognize the mega-element (the literary work itself) since the micro-element and the work concur structurally and thematically (plot), semantically and didactically (symbolism). For instance in *Bruges-La-Morte*, Viane, the protagonist is mourning over the loss of his wife but, simultaneously, he has a love affair with Jane, an actress. He lived in Bruges, a puritan city in which people rebuked such a behavior considered insincere, hypocritical, and immoral. They observed what he was doing through mirrors, lock holes, etc . . . Brugians, that is, citizens from Bruges, constantly spied on Viane and Jane through holes, lock holes, and mirrors to see what he was doing because having a love affair while mourning for one's late wife is scandalous and immoral and they needed to see how far Viane had gone in performing this scandalous act.

The mise en abyme is one of the techniques used to deconstruct the meaning by dint of reconstructing the plot, which finally amounts to delay, shrink and even deconstruct it. Such a process (deconstruction/ delay/shrinking), when reiterated, eventually deconstructs the meaning. Consequently, the signified is de-emphasized at the expense of the signifier because the plot fades away. The deconstruction is precisely one of the trends inherent in modern literature and especially the Nouveau Roman Movement. Most modern novels and those of the nouveau roman in particular bear the seal of deconstruction (*L'Amant* by Duras, *Le Voyeur*, *La Jalousie*, *Le Miroir qui revient* by Robbe-Grillet, to mention but a few), That is why Jean Ricardou who was one of the authoritarian theorists of the Nouveau Roman Movement stated: "Le roman n'est plus l'écriture d'une aventure mais l'aventure d'une écriture" (*Les Nouveaux Problèmes du Nouveau Roman*, 63), which means: "the novel is no longer the writing of an adventure but the adventure of a writing". Deconstruction will reach its peak with Post-Structuralism and its tenors: Derrida, Deleuze, and Foucault.

Many critics, writers have saluted the outstanding impact of symbolists on modern literature. Among these experts' laudation, we can mention this special tribute vicariously paid to Mallarmé by Huymans through Des Esseintes in the section of *A Rebours* where he (Des Esseintes) describes his fervor-infused enthusiasm for the poet: "These were Mallarmé's masterpieces and also ranked among the masterpieces of prose poetry, for they combined a style so magnificently that in itself it was as soothing as a melancholy incantation, an intoxicating melody, with irresistibly suggestive thoughts, the soul-throbs of a sensitive artist whose quivering nerves vibrate with an intensity that fills you with a painful ecstasy." (198, Robert Baldick's translation)

Additionally, as a leading French symbolist poet, his work has generated several revolutionary and artistic schools of the early twentieth century and especially: Dadaism, Surrealism, Cubism, Impressionism, and Futurism. However, one of the most refined qualities of his work stems from three joint areas: an exquisite mastery of prosody; a profound understanding of music; and an unprecedented exploitation of language and phonetics to infuse his poetry with music (use of obsolete, formal, or classical vocabulary, reformulation of semantics, linguistic catharsis, rearrangement of paradigmatic and syntagmatic relationships within a verse structure (proliferation of inversions, *hyperbaton, *anacoluthon, *asyndeton, condensation of the verse structure to avoid narration, description and any kind of lexical development) in order to produce very special musical effects). That is precisely why most experts believe that it is extremely difficult to translate his poetry into English. De facto, the difficulty is due in part to the complex, *polysemic and multilayered nature of much of his work, but also to the important role that the sound of words, rather than their meaning, plays in his poetry. When recited in French, his poems allow alternative meanings which are not a matter of course on reading the work on the page. For example, Mallarmé's *Sonnet en '—yx'* called 'Ptyx' opens with the phrase *ses purs ongles* ('her pure nails'), whose first syllables when spoken aloud sound very similar to the words *c'est pur son* ('it's pure sound'). In the same poem he uses a technique called *"paronomasia". It consists in conjoining words having very similar sonorities to generate imitative harmony or special effects ('sonore' and 's'honore'): "Aboli bibelot d'inanite sonore (voiced) /[. . .] / Avec ce seul objet dont le Néant s'honore" (is honored). In these two verses 'sonore' and s'honore are homophones, that is, words pronounced exactly the same way but spelled differently (homo in Greek = same, and phone = sound). De facto, the 'pure sound' aspect of his poetry has been the subject of musical analysis and has inspired musical compositions. These phonetic ambiguities are very difficult to reproduce in a translation which must be faithful to the meaning of the words.

Additionally, sometimes, Mallarmé blended his musical poetry with *concrete poetry or shape poetry, that is a form of poetry in which the typographical arrangement of words is as important in conveying the intended effect as the conventional elements of the poem, such as meaning of words, rhythm, rhyme and so on. It is sometimes referred to as 'visual poetry', a term that has evolved to have distinct meaning of its own, but which shares the distinction of being a form of poetry in which the visual elements are as important as the text. We prefer to call it holistic, total or gestalt poetry because it synthesizes: poetry, music, drawing, calligraphy, and painting.

In the French review, Vol. 73, No 3, February 2000, Adelia Williams presents a pertinent examination of French poets's contribution to modern arts and literature. In *Verbal Meets Visual: An overview of poésie critique at the fin-de-siècle* she states:

> Through the works of Surrealists, to the efforts of poets of the postwar generation, and up to the present day, French poets have penned hundreds of texts, both in poetry and in prose, that treat the visual arts. Called "*poésie critique*", the genre has spanned our century with ever-increasing dynamism and diversity [. . .] Twentieth century *poésie critique* is both an outstanding and infinitely rich literary genre, as well as a sound contribution to the discipline of art history [. . .] Two essential nineteenth-century "pictorialist" poets: Charles Baudelaire and Stéphane Mallarmé figured also among the foremost art critics of their time. Baudelaire inaugurated the modern poet's preoccupation with the arts by documenting the *Salons* (1845, 1846, 1859). Mallarmé adopted visual elements and strategies in order to approximate some of the qualities of the visible image in poetry. In *Un Coup de dés jamais n'abolira le hasard* of 1897, Mallarmé borrowed the spatial qualities of the plastic arts to create a three-dimensional poem. The innovation of Baudelaire and Mallarmé led the way for our [. . .] century to modern literature (492-493, *Verbal Meets Visual: An overview of poésie critique at the fin-de-siècle*)

Such is the contribution of symbolism to modern literature. Symbolism has also ignited cognitive, therapeutic, and scientific interests. What are their specific manifestations?

5—Cognitive and therapeutic interest

Studying symbolism has become a two-pronged opportunity:

- To achieve the conjunction of arts and neuroscience, which has led to create neuroesthetics, a field of expertise combining neuroscience and arts;
- And, as a corollary of the above, to examine the interests of synesthesia and see how it has enabled us to know three main areas: synesthetes, synesthesia, and the human brain/a new perception of reality/universe. Neuroscience has taught us how the human brain functions through synesthesia. Neurons and synapses merge and cause the merging of the psychic areas to which they belong, adjacent areas are crossed wired (case of color-grapheme synesthesia) which leads to synesthesia/poetry. Neurons and synapses convey information under the form of neurotransmitter chemicals to the thalamus that converts it into poetry/synesthesia. Knowing the function of the human brain through synesthesia implicitly stimulates a therapeutic interest because it also leads to find how certain brain-related diseases occur and can be treated (epilepsy, hemiplegia, etc . . .); to develop certain areas of the brain to render them more effective (memory through the development of the limbic system (hippocampus for instance)); to know the impact of music, sounds, and deep breathings on the brain; to understand how the brain controls the human body, its cognitive aspects (hemispheres, brain-stem, the four lobes of the cerebral cortex, angular gyrus (organ important in the development and understanding of metaphors), the role of emotion in creativity through the limbic system, etc.

a—Synesthetes

Synesthetes possess a highly developed memory, an eidetic memory. They can very easily memorize a phenomenal amount of items: numbers, math equations, formulae, and names since colors elicit numbers or letters. They have a holistic perception of reality (and universe) since it is viewed, heard, tasted, touched, felt, presented in three dimensions, and manifested idiosyncratically. Additionally, because of their memory and high capacity of visualizing objects, they become very proficient in spelling and conducting abstract thinking or visual thinking helpful in quickly and effectively solving complex problems. That was what Einstein used to solve equations regarding general relativity. Some synesthetes have naturally developed visual thinking, which makes possible for them to intuitively find solutions to specific problems. That was the case of Dr. Richard Feynman who used it as well.(these two cases were reported by Van campen in *Synesthesia, the hidden sense* (86).

Synesthetes are also endowed with the ability to make useful connections very quickly. For this reason, they can see hidden items/things and analyze objects at a deeper level. The following image demonstrates the seamless and flawless logic of one test used by Dr. Ramachandran to demonstrate the factual reality of synesthesia. On the left is the image presented to participants, in which a triangle composed of 2s is embedded among a field of 5s. For non-synesthetes this triangle would be hard to identify (displays were presented for one second). However, for synesthetes someone who experience **2s** as red and **5s** as green, the triangle should be more easily identified. **These two images show that synesthetes can perceive hidden items (red numbers and triangles: right image), which is not always the case for non-synesthetes.**

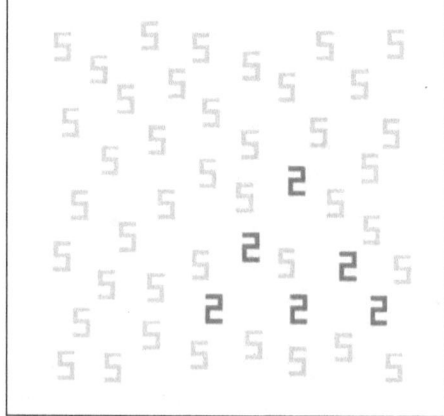

Besides, they seem to be gratified with a kind of sixth sense, an extra sense.

b—Synesthesia and the brain

Synesthesia has enhanced the understanding of the human brain and its correlations. Research from authoritative neuroscientists such as Ramachandran and Hubbard have proven that raw poetry is already encrypted, built-in in specific brain regions (angular gyrus) and pre-disposed to be released when emotion is triggered via the limbic system, translated by the thalamus, and refined into a highly sophisticated language, at the end of which it emerges as cross-modal metaphors, that is synesthesia, a highly polished form of poetry. Indeed, subtle connections are made possible through the angular gyrus, a ridge of the neocortex in the parietal lobe, concerned with the position of the body in space and linking sounds and meaning. This organ is instrumental in the development and understanding of metaphors, making connections between areas apparently unrelated. Precisely, since metaphors are the most poetic figures in literature, it appears that the angular gyrus corroborates the possibility for the brain to generate poetry under specific conditions. Moreover, the theory of brain synaptic plasticity and that of adjacency, now factually scientific, have proven that some forms and modes of synesthesia are commoner than others: color grapheme synesthesia, visual auditory mode (adjacency principle). They have also shown that the brain can change in terms of structure and function (synaptic plasticity principle). Besides, scientists like Gershon and especially Lobsang Rampa have also proven that we use just a small percentage of our mind/brain potential: ten per cent, but if we can increase it beyond this percentage we will factually become genii. That is precisely what Rampa showed in *Chapter of Life, Beyond the Tenth*, and *You Forever*. In the latter, he emphasizes the ability for human beings to increase their brain potential. He declares: "The mind can give us all that we ask if we will let it. There are immense powers latent within the subconscious. Unfortunately most people are not taught how to contact the subconscious. We function at one-tenth consciousness, and-at most-one tenth of abilities. By aligning the subconscious on our side we can achieve miracles." *(You Forever, 206).*

Consequently, studies and findings on synesthesia open new horizons for new forms of knowledge that can be useful in arts and sciences: bridging gaps between sciences and literature/semiotics (neuroesthetics); enabling correlations between thought and language, between sensory modalities and emotions, elucidating cognitive processes, creating new forms of sophisticated arts by means of synesthesia and the assessment of brain potential. That is precisely one of the main goals of neuroesthetics. Through it, scientists and experts analyze the relationships between the brain functions and production of arts/creativity. They study the activity of the brainwaves that is going on while we think, feel, speak or are moved. Through such studies, they have come to realize that the psychic areas of human brain function holistically when we are babies, which might prove that all humans are born synesthetes. In the **"booba-kiki experiment"** Ramachandran subscribes to this possibility but believes that humans lose or keep this disposition over time. The following is an idea of the booba-kiki experiment first designed by Wolfgang Köhler.

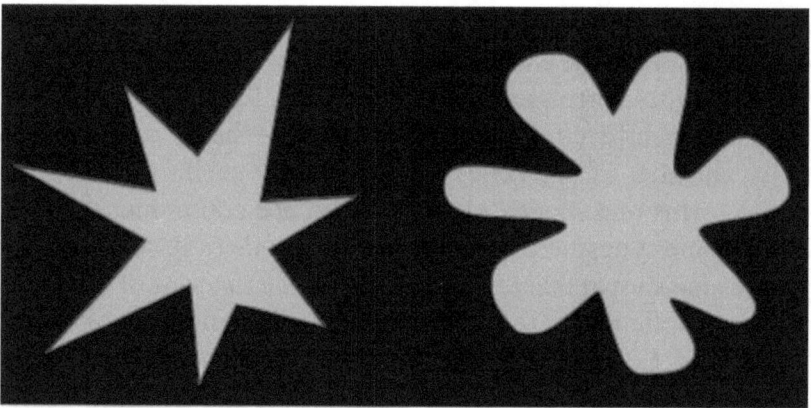

Description: In a psychological experiment first designed by Wolfgang Köhler, people are asked to choose which of these shapes is named **Booba** and which is named **Kiki**. 95% to 98% of people choose Kiki for the angular shape and Booba for the rounded shape.

It is thought that this has implications for language development, in that the naming of objects is not completely arbitrary. The rounded shape may most commonly be named Booba because the mouth makes a more rounded shape to produce that sound. Similarly a more taut, angular mouth shape is needed to make the sound Kiki. The sounds of a K are harder and more forceful than those of a B, as well.

Note also that, in the Roman alphabet, the angular shape mimics the angular letters K and I, while the rounded shape mimics the rounded letters B and O.

The kiki visual shape has a sharp inflection and the sound 'kiki' represented in your auditory cortex, in the hearing centers of your brain, also has a sharp sudden inflection. Your brain performs a cross-modal synesthetic abstraction, recognizing that common property of jaggedness, extracting it, and so reaching the conclusion that they are both kiki.—V. S. Ramachandran

We, humans, can lose the synesthesia acquisition due to the synaptic-pruning process or keep it if this process does not occur. In *The Emergence of the Human Mind* Ramachandran and Hubbard explain the causes and development of synesthesia. They state that:

> The cause of synesthesia can be reduced to two major hypotheses. The first is that synesthesia occurs because of cross-wiring between contiguous brain areas, either because of a deficiency of the proper synaptic pruning that would separate the areas or because of an excess of connections due to other reasons. The second hypothesis promotes a **disinhibited** feedback theory, the idea that there is excess activity between the sensory hierarchy or concurrent pathways because of a disinhibition of the feedback signals that would normally occur. These differences can thus come down to a structural cause, i.e., the extra connections of the first hypothesis, or a functional cause, i.e., the disinhibition of normally existing connections. Strong evidence for both theories makes the debate between the two that much more intriguing.
>
> The cross-wiring hypothesis is based on the idea that adjacent brain areas do not get separated due to a mutation in the gene that would normally regulate the synaptic pruning process.

In this way, the cross-activation is due to "extra wires," to neural connections that were not pruned away. These crossed wires would then lead to a concurrent sense being experienced upon the occurrence of a separate inducer. There is plenty of evidence for the idea that young humans and other organisms have short-lived synaptic connections that are eventually pruned away with maturation. In an infant, the many connections make its cortex barely able to function, proven by low levels of blood flow and bad results at behavioral marker tasks (5). The pruning is a necessary process to allow normal functioning within the world, not necessarily a hampering of ability.

In the adult brain, each sensory area is focused on processing information from a different sensory modality. As stated earlier, studies have shown that synesthetes have activation in both of the cortical areas that are eliciting responses, e.g., for someone who hears words as colored, both the auditory cortex and the visual cortex are activated upon listening to someone speaking. The foundation of the cross-activation theory is that these sensory cortical areas are not initially as specialized as they become later in life. The synapses are pruned over maturation in an experience-dependent manner, and the over-wiring in infants that results in transient connections between adjacent brain areas suggests that these connections are in fact functional before they are pruned. Genetic evidence could lend credibility to the pruning hypothesis. Synesthesia has been demonstrated to run in families and also to occur more in women than in men. This evidence would suggest that there is a chromosomal component to synesthesia, and that it is X-linked because of its prevalence in women. A genetic basis would neatly explain the pruning hypothesis, since a mutation of the gene that initiates the normal pruning process would have the direct result of extra synapses between cortical areas. The ratio in women to men has been found to be 6:1, although there is recent data challenging that. Regardless of that, the solid evidence for familial trends of synesthesia is a telling sign of its heritability factor (152-153).

Other scientists Baron-Cohen (Baron-Cohen, Simon. *Synaesthesia: Classical and Contemporary Readings*) and Richard Cytowic (*Synesthesia: A Union of the Senses*) also believe in the neo-natal and genetic theory of synesthesia (theory based on chromosome legacy structure > karyotype (of men (XY) and women (XX)). This theory shows there are more women synesthetes than men since the synesthetic trait is transmitted through the chromosome X.

Additionally and more importantly, synesthesia can help in thinking on how to treat certain diseases inherent in the brain: epilepsy, hemiplegia. Studies have shown that epilepsy for instance is a brain disorder induced by many factors: brain abscess, brain injury, brain abnormality inherited at birth, brain lack of oxygen, metabolic conditions (*hypoglycemia, *hyperglycemia, low calcium, low or high sodium), etc . . . Generally, these factors result in an uncontrolled excessive and synchronous electrical activity in the brainwaves and bring about the merging of different psychic areas and, sometimes, a consecutive pathological case of transient (temporary) synesthesia. Thus, based on that, researchers might ponder on how to reverse or readjust this amount of electricity in order to treat epilepsy and a number of brain-malfunctions and Traumas. In this picture we see the brain region (red dots) affected by an excessive electrical activity, which will result in epilepsy or seizures (electrical storms).

Illustrations by Dr. Gregory A. Worrell, in *Epilepsy and Neurophysiology*

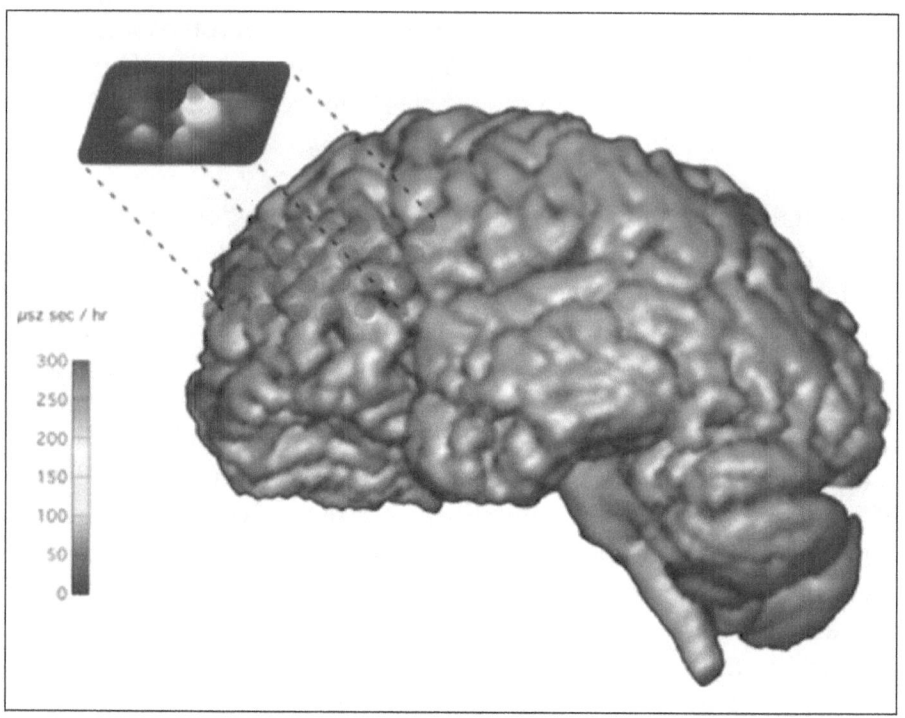

Epilepsy is a chronic medical condition produced by temporary changes in the electrical function of the brain, causing seizures, which affect awareness, movement, or sensation. Epilepsy is a combination of seizures caused by "spontaneously recurrent seizures in which too many brain cells become excited at the same time" (Dr. Robert S. Fisher, M.D. Ph.D., Maslah Saul, M.D. Professor at Standford University, Editorial Board, Epilepsy.com)

More exactly, seizures occur when clusters of nerve cells in the brain, that is, neurons, signal or communicate with each other abnormally. During a seizure, the neurons' normal pattern of activity is disturbed, causing them to fire as many as 500 times per second (normal rate is about 80 times per second). This can cause strange sensations, emotions, and behavior, or convulsions, muscle spasms, and/or loss of consciousness, and, sometimes, synesthesia.

Therapeutic methods can be found through synesthesia

With respect to hemiplegia, its etymology illuminates its meaning and provides a therapeutic hint. The Greek prefix 'hemi' means "half" and the stem "plêgê" refers to "stroke, paralysis". Thus, hemiplegia is a paralysis of one side of the body induced by a lesion of the opposite hemisphere of the brain. It functions crosswise, that is if the left hemisphere is affected, the right side of the body becomes paralyzed and vice versa. Since the left hemisphere regulates the linguistic function, if a patient's left hemisphere is affected, he/she can become dumb or at least have problems relating to language such as *dyslexia, *anomia, transient or irreversible *aphasia. Hemiplegia can lead to accidental synesthesia. In this case, if one knows its specific causes, it is necessary to supply specific psychic areas of the brain (that are affected and provisionally merged) with oxygen because they need it to function adequately and effectively. This oxygen can trigger the proportional redistribution of the flow of electricity/electromagnetism throughout the affected psychic areas and disentangle them, which will re-energize them and make the patient recover quickly. As far as epilepsy is concerned, it is also a brain disease. Neurons communicate each other by chemical and electrical signals through synapses in a process known as synaptic transmission. The fundamental process that triggers synaptic transmission is the action potential, a propagating electrical signal that is generated by exploiting the electrically excitable membrane of the neuron. This is also known as a wave of depolarization. Neurons electrically and chemically communicate with each other through synapses. Such communication is normally regulated by an appropriate amount of electricity but when the amount is very high (rate of 500 times per second), this results in a brain storm known as epilepsy. The following diagram shows communication through neurons in nerve tissue.

NEURONS IN NERVE TISSUE

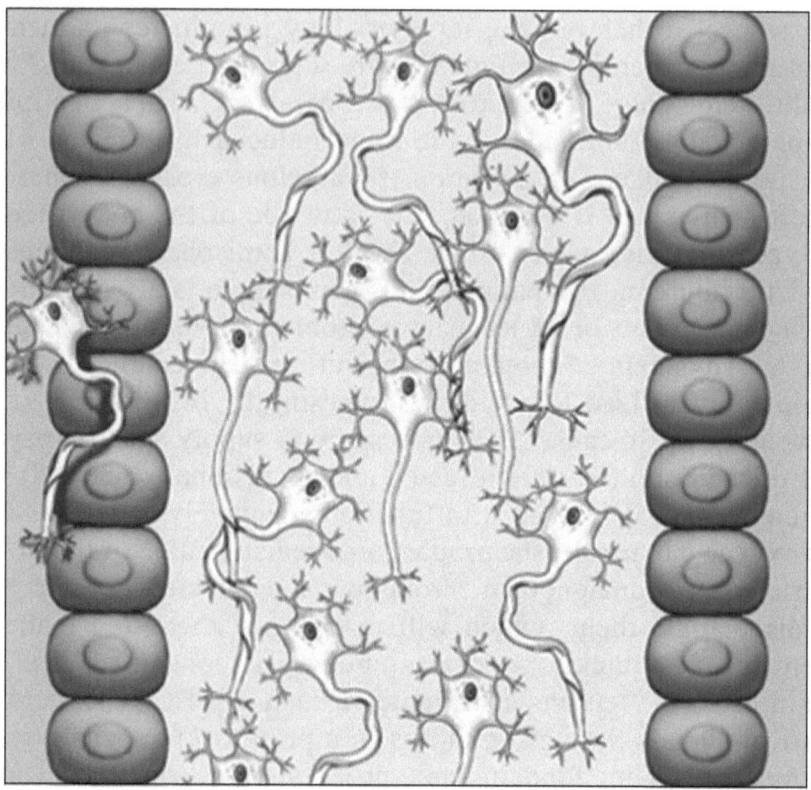

References/Sources: Berkow R. The Merck Manual of Medical Information. 17th ed. New York, NY: Simon and Schuster; 2000 Last reviewed February 2009 by Rimas Lukas, MD

How does epilepsy affect the brain? Etiology and therapy

Patients with epilepsy have seizures caused by unusual electrical activity in part or all of the brain. Doctors can use an EEG to measure this activity, and diagnose epilepsy (as seizures can have other causes). Unusual bursts of electric activity can sometimes be detected between seizures, and are called 'spikes'. The location of these spikes in the brain can help doctors decide what type of epilepsy the person has, and so what treatment to use. In some cases, the condition of epilepsy and hemiplegia can bring about lack of oxygen in the brain. Then, it is necessary to correct this deficiency. Oxygen can be provided by special and specific types of breathing: deep breathing, diaphragmatic breathing, to mention but a few. Actually, the exact cause of hemiplegia is not known in all cases, but it appears that if the brain is deprived of oxygen, this can result in the death of neurons. When the *corticospinal tract is damaged, the injury is usually manifested on the opposite side of the body. For example if one has an injury on the left side of the brain, the paralysis will be on the right side of the body and this can lead to an artificial merging of two or more psychic areas and synesthesia. In this case, it is necessary to provide the brain with oxygen by using special breathing techniques.

Besides, there is a strong correlation between deep breathing with lower abdomen and the brain. Sensei Kanazawa, a world-class researcher in karate, deep breathing, yoga, tai-chi-chuan (he holds the highest level in karate: 10th Dan, and tai-chi-chuan) has methodically conducted research on breathing, Karate and the brain. He has found out that there is actually an interesting correlation between deep-breathing (low diaphragmatic breathing), lower abdomen (hara/seika tanden in Japanese language), and the brain. The following is an account of his findings:

> "It has medically been proven that, after life first appeared and over the course of human evolution, the nerve cells of the intestines and the nerve cells of the brain originally shared a very close relationship.
>
> The role of the stomach is to digest food, and the intestine serves to absorb nutrients from the food we eat. Some 100 billion nerve cells reside in the intestine, approximately one-half of the nerve cells located outside of the brain. Mankind has reached its present form following a lengthy evolution process, and the nerve cells of the intestinal canal and the nerve cells of the brain stem (which is responsible for our life-support functions and comprises the medulla oblongata, the mid brain, the pons, and the thalamus and the hypothalamus) are essentially the same. A portion of the nerve cells from the intestinal canal migrated upward and achieved further development in the head, combining with the cerebrum to form the brain and facilitate human intelligence [. . .] The continuous practice of deep-breathing and karate not only

> leads to increased concentartion but also seves as a form of intellectual training since the nerve cells of the intestinal canal and those of the brain are essentially the same[. . .] Training the low abdomen leads to greater concentration and endurance, and strengthening the hara strengthens the mind and the brain (16-17, *Black Belt Karate*).

In light of this finding, it can be inferred that deep-breathings, that is, those involving the lower abdomen executed by lowering the diaphragm, inhaling and exhaling (inhalation and exhalation having the same length) very slowly and deeply, can positively impact the brain and improve health. The correlation of the brain and the stomach has also been validated by a renown scientist, Dr. Michael Gershon, a neurobiologist at New York's Columbia-Presbyterian Medical Center. Gershon recently explained to *Psychology Today* (PT) how an independent network of over 100 billion neurons in the gut not only signals our bodies to stress but causes illness. He is the author of *The Second Brain* (HarperCollins, 1999). He states that human stomach has neurotransmitters similar to the brain. The stomach is factually our second brain. He explains how the stomach communicates with the brain. The following is an interview he granted to *Psychology Today (PT) and published* by PT Staff on May 01, 1999. This interview concur with Sensei Kanazawa's research. Therefore, it confirms the correlation between the brain and the stomach considered as the "second brain"

Q Why do we need a second brain?

A Most importantly, to control digestion. It also works with the immune system to protect us from hostile bacteria.

Q Does it use neurotransmitters?

A Actually, 95% of all *serotonin in the body is in the gut, where it triggers digestion. Nerve cells in the gut also use serotonin to signal back to the brain. This information can train us not to eat certain foods by communicating pain, gas and other terrible feelings.

Q Does the brain in our heads influence the "second brain"?

A Yes. Butterflies in the stomach arise when the brain sends a message of anxiety to the gut, which sends messages back to the brain that it's unhappy. But the gut can also work in isolation. Have you ever gotten a gut feeling about someone, or are anxious about butterflies in your stomach? That's because you have a second brain in your bowel.

Q How does this brain influence irritable bowel syndrome (IBS), which many believe is a psychological problem?

A Irritable bowel syndrome, whose symptoms include abdominal pain accompanied by loose stool, affects 20% of Americans. But doctors often dismiss its severity, attributing IBS to psychoneurosis because they don't know exactly what it is. I propose that the second brain is the cause. Antidepressants like SSRIs (Selective Serotonin Re-uptake Inhibitors or Serotonin-Specific Reuptake Inhibitor is used in the treatment of depression, anxiety disorders or some personality disorders), when used in doses too low to treat depression, are effective immediately in IBS patients. Prozac takes weeks to kick in. This suggests that the drugs work not on the brains of people with IBS, but in the bowel.

It is noteworthy that breathing techniques fall under the category of natural therapy including natural forms of treatment such as "ki exercises" conceived to re-energize the brain. In fact, "Ki" in Japanese language means energy, but it is not specific to Japanese and can be equated with "prana" (meaning energy, breath, life-force in Sanskrit). "Ki" exercises are based on breathing and have a scientific application and explanation. Breathing, stricto sensu, is at the very root of life. In Sanskrit and several languages **breath** refers **to life itself.** In Latin, for instance, the word **"anima, ae"** means **soul, life, breath, air** indistinctively because to have air presupposes to have energy, a soul and, therefore, life. The human brain and its neurons need oxygen to function appropriately and oxygen is predominantly provided by air and a cosmic essence. Both of them are channeled through proper breathing. It is scientifically proven that if neurons remained deprived of oxygen for just a few minutes, death will occur. Therefore, by performing special types of exercises; "total and full breathing", performed by lowering the diaphragm while keeping the spinal cord erect, inhaling and exhaling slowly and deeply while being completely relaxed, one can stimulate the **pineal gland** in the brain. This will fortify/reenergize the brain and proportionally redistribute the amount of electricity and electromagnetic energy within it and, accordingly, treat diseases such as epilepsy and hemiplegia if they are caused by lack of oxygen. However, it is indispensable to know the precise cause (s) of the disease to apply the appropriate therapy. "Ki exercises" can provide massage and oxygen to the brain and thus act as a kind of brain massotherapy. An article titled *Hope and the Pineal Rhythmic Breathing and Chanting* written by Gary Vey perfectly fits in the cardinal function of breathing and its positive impact on the pineal gland. This gland is very important and located near the thalamus. It secretes melatonin, the hormone that regulates the circadian rhythms including the awake-sleep cycle. Gary Vey declares:

> The pineal gland sits on the roof of the 3rd ventricle of the brain, directly behind the root of the nose and floats in a small lake of cerebrospinal fluid. It doesn't have a blood-brain barrier like other brain structures, where certain molecules in the circulating blood are blocked. Instead it relies on a constant supply of blood delivered through a particularly rich vascular network. Activities such as rhythmic breathing and chanting create an oxygen rich supply to the pineal gland. *Yogic practices/rhythmic breathings for 3 months resulted in an improvement in cardiorespiratory performance and psychological profile. The plasma melatonin also showed an increase after three months of yogic practices. The systolic blood pressure, diastolic blood pressure, mean arterial pressure, and orthostatic tolerance did not show any significant correlation with plasma melatonin. However, the maximum night time melatonin levels in yoga group showed a significant correlation ($r = 0.71$, $p < 0.05$) with* well-being score[...]These observations suggest that yogic practices can be used as psychophysiologic stimuli to increase endogenous secretion of melatonin, which, in turn, might be responsible *for improved sense of well-being.*

Serotonin is a neurotransmitter found in your gastrointestinal tract, platelets and central nervous system. Both melatonin and serotonin play several roles in human behavior and cognition. Serotonin, in particular, has some effects on cognitive functions including memory and learning. Regarding DMT, it is substance, a drug supposedly made with melatonin. These three elements can be properly channeled and stimulated by specific breathings/yoga practices to positively impact the brain and significantly improve health. The following are the formulae of serotonin, melatonin and DMT with their specific components:

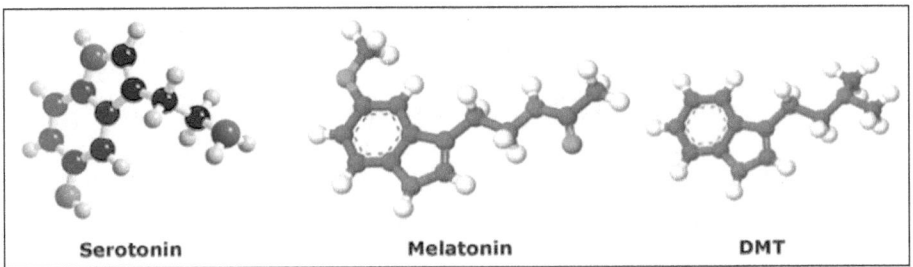

| Serotonin | Melatonin | DMT |

These are additional reflections of the author with respect to the pineal gland, serotonin, melatonin and DMT

> More dramatic evidence comes from mystical and "near death experiences" induced by a chemical called Dimethyltryptamine (DMT). The traditional "sacrament" of South American shamans, DMT closely resembles both melatonin and seratonin and occurs naturally in some tropical vegetation as well as the human body. It is thought that stored DMT is released from the pineal gland just prior to death, causing the out-of-body experience and mystical visions reported by NDE survivors. While there has been no empiricial evidence linking the pineal gland to the production of DMT, its association with melatonin (from which it is made) seems to strongly suggest this possibility.

It is also possible to lengthen one's span of life through ki exercises. Within the same therapeutic framework, meditation can be used to treat the brain. It transcends the use of the five senses and pertinently fits in cases of synesthesia/five senses. In an article titled *Beyond the Five senses: the Healing Power of Meditation*, Dr. Deepak Chopra elaborates on this particular aspect. He states:

> The body isn't a sloshing bag of biochemicals; it's a field of intelligence or consciousness that is constantly responding to your thoughts, emotions, and the input of sensory impressions.

By immersing yourself in nourishing sounds, touch, sights, tastes, and smells you can enhance the mind's ability to heal and balance the body. To go even deeper and directly access the mind's infinite healing powers, the most effective tool of meditation [. . .] A large body of scientific research has established that having a regular meditation practice produces tangible benefits including: more efficient oxygen use by the body/brain; improve immune function; lower blood pressure and hypertension, slower heart rate and decreased cholesteol levels [. . .] **(58, 59).** We must add that meditation is also a form of mental cleansing, a bath taken internally.

Moreover, using specific sounds in a methodic fashion can turn out to be a form of therapy for the brain. Indeed, the power of sounds for healing the brain is technically possible when taking into account the scientific nature of sound and that of the brain. The sound can be defined as a sequence of waves of pressure/a vibratory phenomenon that propagates through flexible media such as air or water. It has properties. Sound waves are often simplified to a description in terms of sinusoidal plane waves, which are characterized by generic properties. The most important ones are:

- **Frequency f, or its inverse, the period**
- **Period = T = $\frac{1}{f}$**
- **Wavelength: λ Wave speed: f λ -> product of frequency and wavelength**
- **Momentum (P): mass (m) x velocity**
- **Sound pressure** or **acoustic pressure,** that is the local pressure deviation from the ambient (average, or equilibrium) atmospheric pressure caused by a sound wave
- **Sound pressure level** (SPL) or **sound level** is a logarithmic measure of the effective sound pressure of a sound relative to a reference value. It is measured in decibels (dB) above a standard reference level.

With respect to the brain, it is the center of the nervous system, the most sophisticated device of the human body. It weighs three (3) pounds composed of 100 billion neurons. It is made up of matter: heavily folded muscles with five (5) major parts: the cerebrum, the cerebellum, the brain stem, the spinal cord (which is the extension of the nervous system), neurons (brain cells) and synapses. By virtue of the synaptic and brain plasticity principle, we know that the brain is flexible. Thus, it can change in terms of structure and function (in previous chapters, we showed that synapses pertaining to different psychic areas can tangle and cause the tangling and merging of different psychic areas, which this leads to synesthesia). From experience, it has been proven that sound can impact matter, affect it negatively or positively because it is endowed with properties that enable it to reach this goal: a frequency, a period, a wavelength, an intensity, a momentum if we consider that the sound has mass because the equation of the momentum (P) is: mass (m) multiplied by velocity (v): **P = m x v**. Sound is composed of waves of pressure and can damage objects, affect their structure and function. In *You for ever* Dr. Rampa illustrates the power of waves of pressure but he uses the term 'vibrations' rather than 'waves of pressure'. He states:

> "We all know the very simple illustration of the power of vibrations; soldiers who are marching along keeping step will break back that step and walk across a big bridge in any disordered array of paces. The bridge may be capable of withstanding the heaviest mechanical. It may be capable of bearing a whole succession of armored tanks rattling across, or it may bear a whole load of railway locomotives, and it will not deviate more than its designated amount through that load. Yet let a column of men march in step across that bridge, and it will set up a **momentum** that causes the bridge to sway and bounce, and eventually to collapse. Another illustration we might give in the matter of vibration is that of a violinist; if he takes his violin he can, by playing **a single note for some seconds,** cause vibrations to build up in a wine glass with the result that the glass will shatter with a surprisingly loud explosion. The soldiers are one end of our illustration on vibration, and the other end? Let us consider Om. If one can say the words, OM MANI PADME HUM **in a certain way** and keep on saying that for a few minutes, one can build up a vibration of quite fantastic strength. (231-232)

In light of these considerations, it appears that the sound or set of vibrations can be endowed with a tremendous power and affect matter if it is applied to it with the relevant frequency or momentum (if we consider that sound waves have mass). Since frequency (f) is, by definition, the number of vibrations or waves passing a point per second measured in hertz (Hz (S^{-1})), in Dr. Rampa's quote **"a single note for some seconds"** stands for the frequency of the note; **"a momentum"** stands for the frequency of the sound wave or vibrations set up by the march of soldiers, **"in a certain way"** is a periphrastic construction standing for the appropriate frequency. It follows that, by virtue of laws of physics, the sound or vibrations applied with the appropriate frequency (a single note for some seconds) can break a glass or a bridge and cause it to collapse. Now, given that a glass, a bridge, and the human brain, are fundamentally composed of matter, what applies to the former (glass and bridge) can also apply or be extended to the latter, that is, the human brain. Additionally, studies on synesthesia have enabled a better understanding of the brain. They have proven that the brain is malleable, flexible (brain/synaptic plasticity principle). Accordingly, sounds can be used to affect the brain and change it structurally or functionally, positively or negatively. De facto, sounds can be used to treat patients suffering from certain brain diseases under the condition that the causes of such diseases be known. This treatment fall within the range of phonotherapy and musicotherapy. Phonotherapy, from the Greek 'phone' meaning 'sound' and 'therape' treatment, is the art of treating patients by means of specific sounds such as mantras. "OM MANI PADME HUM" that Dr. Rampa mentioned in his reflection is a mantra: a special combination of sounds based on a very specific frequency note designed for enhancing/supporting concentration, meditation/or/and prayer. The power of sounds is strongly validated by several researchers. They buttress Dr. Rampa's reflection. One of them, Nozedar Adele, in The Illustrated Signs and Symbols Sourcebook states: "Everything in the universe has its own frequency, its own vibration. Some frequencies are so powerful that they can destroy physical objects, again a good example is the soprano whose top note is so pure it can shatter a glass [. . .] certain very low frequencies can destroy matter by scrambling molecules. Sound can cause avalanches"(412).

With respect to musicotherapy, it is vicinal to phonotherapy and designates a form of treatment using music as a healing medium. It is an allied health profession and one of the expressive therapies, consisting of an interpersonal process in which a trained music therapist uses music and all of its facets—physical, emotional, mental, social, aesthetic, and spiritual—to help clients to improve or maintain their health. Musicotherapists primarily help clients improve their health across various domains (e.g., cognitive functioning, motor skills, emotional and affective development, behavior and social skills, and quality of life) by using music experiences (e.g., free improvisation, singing, songwriting, listening to and discussing music, moving to music) to achieve treatment goals and objectives. It can change the frequency of brainwaves, setting them at the appropriate rate to reduce emotion or channel it positively. It is considered both an art and a science. It might be interesting to make a quick reference about the historical background of musicotherapy because it underscores its efficacy as a therapy.

Cursory historical background

Music has been used as a healing force for centuries.Music therapy goes back to biblical times, when David played the harp to rid King Saul of a bad spirit. As early as 400 B.C., Hippocrates, Greek father of medicine, played music for his mental patients. Aristotle described music as a force that purified the emotions and assigned it a *cathartic function. In the thirteenth century, Arab hospitals contained music-rooms for the benefit of the patients. In the United States, Native American medicine men often employed chants and dances as a method of healing patients. Music therapy as we know it began in the aftermath of World Wars I and II. Musicians would travel to hospitals, particularly in the United Kingdom, and play music for soldiers suffering from war-related emotional and physical trauma. The Turco-Persian psychologist and music theorist Al-Farabi (872-950), known as "Alpharabius" in Europe, dealt with music therapy in his treatise *Meanings of the Intellect*, where he discussed the therapeutic effects of music on the soul. Robert Burton wrote in the 17th century in his classic work, *The Anatomy of Melancholy*, that music and dance were critical in treating mental illness, especially melancholia.

Goals and methodology

Research has proven that music can positively change the brain function. Indeed, one therapy model inspired from neuroscience, called "neurological music therapy" (NMT), is "based on a neuroscience model of music perception and production, and the influence of music on functional changes in non-musical brain and behavior functions." In other words, NMT studies how the brain is without music, how the brain is with music, measure the differences, and use these differences to cause changes in the brain through music that will eventually affect the client non-musically. As one researcher, Dr. Thaut, said in *An Introduction to Music Therapy Theory and Practice*: "The brain that engages in music is changed by engaging in music." (475). Besides, NMT trains motor responses (i.e. tapping foot or fingers, head movement, etc.) to better help clients develop motor skills that help "entrain the timing of muscle activation patterns". (Roth, Edward. *Neurologic Music Therapy. Academy of Neurologica Music Therapists* Retrieved 19 April 2011). Musicotherapy has a special impact on patients with emotional or mood disorders or with the propensity of having them. For that purpose, it can serve as a creative outlet to release or control **emotions** and find ways of coping with difficult situations. Music can improve one's mood by reducing stress and lowering anxiety levels, which can help counteract or prevent depression. It can be used as brainwave rhythmic entrainment (brainwave entrainment or "brainwave synchronization," is any practice that aims to cause brainwave frequencies to fall into a step with a periodic stimulus having a frequency corresponding to the intended brain-state (for example, to induce sleep), usually attempted with the use of specialized software. It purportedly depends upon a "frequency following" response on the assumption that the human brain has a tendency to change its dominant EEG frequency towards the frequency of a dominant external stimulus) for physical rehabilitation in stroke victims and patients with mood disorders. Music has been shown to affect portions of the brain. Part of this therapy is the ability of music to affect emotions and social interactions. Recently, after Colorado shooting in a movie theater, one of the victim, Petra Anderson, who was shot at the head, miraculously recovered from a brain injury and music positively helped her to recover significantly. In an interview scheduled by Anderson Cooper on CNN, Dr. Sanje Gupta, a neurosurgeon and journalist, said: "music can have an amazing effect on the brain. Just hearing or reminding sounds crossing from the left side of the brain to the right side can truly harness the brain". Research by Nayak et al. showed that musicotherapy is associated with a decrease in depression, improved mood, and a reduction in state anxiety. Both descriptive and experimental studies have documented effects of music on quality of life, involvement with the environment, expression of feelings, awareness and responsiveness, positive associations, and socialization. Additionally, Nayak et al. found that musicotherapy had a positive effect on social and behavioral outcomes and showed some encouraging trends with respect to mood. Given the fact that musicotherapy works on emotion to channel it positively and heal patients, it works on brain areas specializing in emotions: the limbic system and more precisely, the amygdalae, hypothalamus, and several others. Emotions are thought to be related to activity in brain areas that direct our attention, motivate our behavior, and determine the significance of what is going on around us. The following brain structures are currently thought to be involved in emotion: the amygdalae, prefrontal cortex, anterior cingulate cortex (ACC), ventral striatum, insula, cerebellum, and the right Hemisphere.

The treatment of epilepsy and stroke

Epilepsy

Research suggests that listening to Mozart's piano sonata K448 can reduce the number of seizures in people with epilepsy. This has been called the "Mozart effect." However, this has not yet reach consensus among scholars. More recent research suggests that music can increase patient's motivation and positive emotions. Current research also suggests that when musicotherapy is used in conjunction with traditional therapy it improves success rates significantly. Therefore, it is hypothesized that music therapy helps stroke victims recover faster and with more success by increasing the patient's positive emotions and motivation, allowing them to be more successful and driven to participate in traditional therapies.

Stroke

Current research shows that when musicotherapy is used in conjunction with traditional therapy it improves rates of recovery and emotional and social deficits resulting from stroke. A study by Jeong & Kim examined the impact of musicotherapy when combined with traditional stroke therapy in a community-based rehabilitation program. Thirty-three stroke survivors were randomized into one of two groups: the experimental group, which combined rhythmic music and specialized rehabilitation movement for eight weeks; and a control group that sought and received traditional therapy. The results of this study showed that participants in the experimental group gained not only more flexibility and wider range of motion, but an increased frequency and quality of social interactions and positive mood.

Music has proven useful in the recovery of motor skills. Rhythmical auditory stimulation in a musical context in combination with traditional gait therapy improved the ability of stroke patients to walk. The study consisted of two treatment conditions, one which received traditional gait therapy and another which received the gait therapy in combination with the rhythmical auditory stimulation. During the rhythmical auditory stimulation, stimulation was played back measure by measure, and was initiated by the patient's heel-strikes. Each condition received fifteen sessions of therapy. The results revealed that the rhythmical auditory stimulation group showed more improvement in stride length, symmetry deviation, walking speed and rollover path length (all indicators for improved walking gait) than the group that received traditional therapy alone.

On the basis of all these considerations, it can be inferred that sounds through phonotherapy and music can have a significant impact on the brain. They can factually be utilized to heal certain brain diseases when their causes are explicit since the etiology of a disease (cause) enable therapists to figure out the appropriate therapy. Consequently, synesthesia has enlightened researchers on the fact that just like sound can act on matter and the brain, phonotherapy and music can act on the brain.

6—Scientific interest

Synesthesia also applies to scientific areas and physics in particular. In fact, it provides food for thought with respect to the field of quantum mechanics. Indeed, since synesthesia shows that our universe is regulated by the principle of interconnectedness from the microcosmic to the macrocosmic levels, it might then become possible to connect the probabilistic world of sub-atomic particles (quantum mechanics) to that deterministic of stars and galaxies (relativity) under the aegis of a theory proposed by the cosmologist Stephen Hawking called "quantum gravity" and used for research on black holes. *Dictionary of Important Theories, Concepts, Beliefs and Thinkers* adds a noteworthy detail: *"Superstring Theory, which seeks to explain subatomic forces in terms that embrace relativity and QUANTUM THEORY, is thought by some to hold the best promise of uniting all four forces in a "theory of everything."* "(378)

Quantum Gravity (QG) is the field of theoretical physics which attempts to develop scientific models that unify quantum mechanics (describing three of the four known fundamental interactions) with general relativity (describing the fourth, gravity). It is hoped that development of such a theory would unify into a single mathematical framework all fundamental interactions and to describe all known observable interactions in the universe, at both subatomic and cosmological scales. Additionally, the interconnectedness principle reflects on the quantum non-locality principle. By virtue of it, atoms/elements can exist at many different places at the same time and something occurring in one place may have a direct impact on something else existing at a farther place. That is what John Bell proposed and proved in: *"Speakable and Unspeakable in Quantum Mechanics" (14)*

There is another direct correlation between synesthesia and physics with respect to the second law of thermodynamics. This law states: "the total entropy of an isolated system increases over time". The word "entropy" was coined in 1865 by a scientist, Rudolph Clausius, to refer to the measure of disorder in isolated systems. Therefore, entropy is chaos that can be found in any isolated system and an aged universe. As a matter of fact, synesthesia lays out the image of a coherent universe, governed by interconnectedness and, as such, highly organized. Since a coherent universe is regulated by high order rather than chaos, it entails a low entropy or a lack of entropy. Hence the fact that synesthesia complies with high order and accordingly, low or zero entropy, which was precisely Baudelaire's conviction. He believed that the beginning of the universe was governed by perfect order/unity, which implied the pre-existence of a supreme architect vouching for such order. On the basis of this supreme order, people used to communicate synesthetically since the world was regulated by an "indivisible totality" and, by the same token, lack of entropy or low entropy. In *Salon de 1850* Baudelaire declared: "It would be surprising if the sound were not able to suggest the color. Likewise, it would be surprising if colors were not able to give an idea of a melody and if the sound and color were not apt to translate ideas. Such is how things have always been expressing themselves by means of a reciprocal analogy since God created the world as a complex and indivisible totality." (16) In light of this law, it can be inferred that all of us were born synesthetes and over time we lost this faculty. As a baby, one has 0 entropy or very low entropy (A baby body can precisely be conceived as a highly ordered system or a coherent universe of cells, atoms, nuclei, electrons, protons, etc . . .), but, as time goes by, entropy increases exponentially and the baby loses synesthesia (increased entropy). However, there is an enigma that needs to be decrypted: why do some people keep synesthesia for their entire life-in spite of entropy increased by time-whereas others lose it? This enigma will be decrypted with the progress of studies on synesthesia. Some scientists such as Ramachandran and Hubbard in particular believe that, over time, the pruning process occurs and sensory perceptions and psychic areas are no longer merged (there is no more "cross-wiring"), which causes the loss of the synesthesia trait (153, *The Emergence of the Human Mind*).

It is noteworthy that another correlation exists between the universe and synesthesia. Actually, the word 'universe' is classically and semantically related to order. Indeed, 'universe' has the same meaning as 'cosmos'. The Greek word 'Kosmos' (κόσμος), which means 'order', is antithetical to the concept of chaos. Today, this word is generally used as a synonym of the universe (considered in its orderly aspect). A Greek myth helps to understand its meaning. It relates that before the universe was created, there was an inform mass regulated by Chaos. Then, there came Logos. However, Chaos, vector of disorder and Logos, vector of order were utterly incompatible by nature and status. Subsequently, due to their antagonistic status, they got involved in a gigantic fight. Finally Logos defeated Chaos and restored order in the Cosmos. That is how Cosmos was born. This myth symbolizes the concept of order as a pre-requisite for coherence and unity in the universe. It suggests that the universe or cosmos was controlled by order at its inception, which gave way to interconnectedness, organization, and zero entropy found in the mechanism and internal function of synesthesia but, again, the question arises: if synesthesia can, to a certain extent, be guaranteed by low entropy or lack entropy, why do some people still have synesthesia despite the flow of time, that is the increase in entropy (since entropy increases exponentially over time)? The answer will be given as researchers secure new findings in the field of synesthesia.

Synesthesia also postulates the existence of the fourth dimension, that is, time. Why is this dimension important? It is important because it can trigger the advancement of knowledge and elucidate several theories that are still enshrouded by mystery in contemporary science: teleportation, invisibility, and magnetic invisibility, time travel, disappearances of peoples in certain locations such as the triangle of Bermuda and Shetland Islands, parallel universes, time travel, Near Death Experiences NDE), etc . . . On page 69 of *Chapters of Life*, Dr. Rampa conducts an interesting discussion on the fourth dimension. Precisely, as a beautiful illustration of the theory of correspondences between the terrestrial and supra-terrestrial, synesthesia can significantly optimize or simplify the understanding of the fourth dimension. The fact of the matter is that, as we have mentioned earlier, all elements are interconnected in the universe and each of them has its correspondence in the material or/and the ethereal (Law of Hermes, and Swedenborg's theory of correspondences, Baudelaire's correspondences). We have plenty of instances of correspondences in the universe such as those found in sensory perceptions: sight, hearing, touch, smell, taste through synesthesia/ neuroscience; in protons and neutrons; quarks and anti-quarks; matter and anti-matter in quantum mechanics; mass and energy summarized by the formula $E = MC^2$, (which shows that mass/matter (M) can be converted into energy (E) at a very high speed, close to that of light (Celerity)), etc . . . . We can extend the range of these correspondences to the infinitesimal scale. In light of this postulate, we consider that the three dimensional universe in which we live also has a correspondence/ extension in the space-time continuum: the fourth dimensional universe, that is, time or space-time because, since Einstein's bold invention of the relativistic equations, we have been cognizant that space-time is a continuum, a flexible substance, or just two interrelated expressions of the same equation. *The Dictionary of Important Theories, Concepts, Beliefs and Thinkers* provides additional explanations on this notion as follows: "In his special theory of relativity Einstein saw space and time as intimately interconnected; instead of three-dimensional space existing in time, he proposed a fourth-dimensional *space-time:* occurrences in the universe cannot be described in terms of space or time alone but only in terms of both at once" (340). In physics, spacetime (or space-time continuum), is any mathematical model that combines space and time into a single continuum. Spacetime is usually interpreted with space as being three-dimensional and time playing the role of a fourth dimension that is of a different sort from the spatial dimensions. From a Euclidean space perspective, the universe has three dimensions of space and one dimension of time. By combining space and time into a single manifold, physicists have significantly simplified a large number of physical theories, as well as described in a more uniform way the workings of the universe at both the supergalactic and subatomic levels.

Consequently, by virtue of the theory of correspondences whose dynamic manifestation is synesthesia, the three-dimensional universe of space has its correspondence: fourth dimensional universe of space-time (because space and time do not exist independently). Precisely, the holistic aspect of synesthesia conceived as an extra sense that enables synesthetes to see/view objects from several angles and several possible perspectives to discover hidden things such as hidden triangles in geometrical figures (see diagrams of hidden triangles with Wolfgang Köhler. and Ramachandran on page 129)—By hypothesizing—grants the technical possibility to the synesthetic brain to see or, at least, intellectually understand the fourth dimension: space-time. It can also be assumed that the sight of synesthetes has a vibration frequency different from that of most human beings. This enables them to see things most individuals cannot: the fourth dimension is one of them. In fact, most humans can see only within a limited range of vibration frequencies, which is called the visible spectrum. According to Dr. Rampa in You forever (36), certain animals like dogs for instance can see beyond the visible spectrum (dogs see the aura). However, with a special training, most of us can extend our limited range beyond the visible spectrum by honing the processing of the brain psycho-visual area. There are special exercises for that purpose. Technically and hypothetically, synesthetes' brains are pre-disposed or endowed with the ability to make them see beyond the limited range of electromagnetic vibration frequencies. This also, theoretically, can confer upon them the ability to see the fourth dimension It is interesting to note that the fourth dimension is not a concept inherent in physics. It also pertains to the field of epistemology and the arts. It is tacitly examined in Plato's *Myth of the Cave* (where it is symbolized by the universe of archetypes or essences). However, synesthesia and neuroscience can provide a better cognitive framework of it. It is artistically expressed in Picasso's cubism and Dali's surrealism (Both of them were influenced by symbolism). In *Hyperspace*, the renown physicist Michio Kaku validates this viewpoint. He declares:

> "Simply put, Cubist art embraced the fourth dimension. Picasso's painting is a splendid example, showing a clear rejection of the perspective, with women's faces simultaneously from several angles. Instead of a single point of view. Picasso's painting show multiple perspectives, as though they were painted by someone from the fourth dimension, able to see all perspectives simultaneously (*Hyperspace*, 65)

One of the best illustrations of this pictorial approach is Dora Maar's portrait by Picasso. Dali also referred to the fourth dimension surrealistically. Dr. Michio Kaku furthered comments on Dali's painting: "Salvador Dali used Hinton's tesseract, an unraveled hypercube, in his famous painting *Christus Hypercubus* on display at the Metropolitan Museum of Art in New York, which depicts Christ being crucified on a four-dimensional cross." (70) The illustration of the fourth dimension by Dali also attests to the fact that artists are visionary, futuristic, and by the same token, can be at the forefront of scientific knowledge. Sometimes, artists, poets, intuitively discover important phenomena that will later be re-discovered and examined by scientists. That was the case of synesthesia since it was first discovered, perceived by artists (poets and philosophers precisely) and later re-discovered and analyzed by scientists. Hence, again the necessity for artists (poets, philosophers, painters, musicians, etc . . .) and scientists to collaborate. They should synergize their heuristic endeavors for the sake of promoting knowledge, which will boost the reign of the human species on earth and the universe. Actually, research on synesthesia can epitomize and catalyze this collaboration.

The following is Dali's Christus Hypercubus on display, depicting Christ crucified on a four-dimensional cube.

Consequently, by pondering upon the connections between physics, neuroscience, Cubism, Surrealism and the fourth dimension, synesthesia posits a deterministic approach of the universe. De facto, since all elements are interconnected at a microcosmic and macrocosmic level, the universe is a *deterministic system, that is, regulated by a causality principle rather than randomness. It follows that something happening in location x can significantly affect what happens in location y, which explains Edward Lorenz's Chaos Theory formulated in these words: "in a chaotic system a tiny change in one part can cause enormous change elsewhere." Lorenz further developed his theory by acknowledging that the chaotic system is controlled by fixed rules, whether physical or mathematical but the behavior is so wildly unpredictable that it looks as if it were caused by chance. One important property of a chaotic system is that a small alteration to the initial values may cause a great change in the long run. This is well-known as the butterfly effect.

SUMMARY OF CORRESPONDENCES & FEATURES OF INTERCONNECTEDNESS

1—Microcosm -< Synesthesia -> Macrocosm
2—Microcosm <-=> Macrocosm,
3—3 Dimensional Universe <=-> 4 Dimensional Universe
(length, width, height) <=> (length, width, height + space-time continuum (see Einstein's general relativity))

This can be read as follows:

1—Synesthesia functions as a bridge connecting the Microcosm (quantum world) and the Macrocosm (super-galactic world). It actually epitomizes this correspondence/connection. It also stands for <u>intra-cosmic</u> connection/s (within the <u>same universe</u>) and <u>inter-cosmic</u> connection/s (<u>between different universes</u>)
2—The Microcosm reflects/corresponds to the Macrocosm (see Plato/Platonism, Baudelaire)
3—The 3 Dimensional Universe corresponds to/ has a projection to the 4 Dimensional Universe

<u>MY HYPOTHESIS & PRACTICAL INTERESTS OF THE 4TH DIMENSIONAL WORLD</u>

The special brain of synesthetes endows them with an extra-sense. This extra-sense is remarkable and suggests the theoretical possibility for synesthetes to see the 4 dimensional universe. Technically and hypothetically, synesthetes' brains are pre-disposed or endowed with the ability to make them see beyond the limited range of electromagnetic vibration frequencies (visible spectrum). This also, theoretically, can confer upon them the ability to see the fourth dimension. Systematic studies focused on this possibility might help researchers to scientifically understand the fourth dimensional universe and solve problems, enigmas such as: teleportation, invisibility, magnetic invisibility, disappearances of persons in the Triangle of Bermuda and certain other areas on earth, parallel universes, time travel, scientific explanations of near death experiences (NDE), and building tunnels/short-cuts in the fabric of space-time continuum (wormholes) to undertake inter-galactic trips—that way a journey supposed to last one thousand years from point A of our galaxy to point B of another galaxy (or point A of our galaxy to point B of our galaxy) will take one or a few days to complete. Consequently, the scientific understanding of the fourth dimension will definitely trigger scientific and technological revolutions and positively change our lives on earth. My hypothesis stems from the fact that the fourth dimension fundamentally deals with time or more exactly, space-time (Einstein has taught us that both are interdependent, monolithically united). Now, time is a limitation for us and Einstein has proven that nothing cannot exceed the speed of light (300,000 kilometers per secundum), which is not absolutely true since the speed of thought exceeds that of light. However, he has taught us something much more important: when we travel at a speed closer to that of light, time goes very slowly and time travel becomes possible. It follows that if we can thoroughly understand the scientific aspect of the fourth dimension, we will likewise understand that of time, given that the fourth dimension focuses on time. Therefore, we will use this knowledge to "master" the time and travel extremely fast and, equally, cover extremely far distances. Finally, through a meticulous study and understanding of the extra-sense inherent in synesthesia and synesthetes, we can kill "several birds" (Time travel, intergalactic travel, invisibility, magnetic invisibility, teleportation, building wormholes, solving problems regarding inexplicable disappearances of peoples in certain locations on earth, understanding the mystery of death through NDE phenomena, etc...) with one "scientific/rational stone". Nevertheless, to complete these achievements we will need a tremendous amount of technological support. Consequently, understanding the fourth dimension and synesthesia is very important if we, humans, aspire to be the lords of the universe. My hypothesis sounds like science fiction or, at best, an epistemological curiosity, but it is neither science-fiction nor such a curiosity! It is science hypothesis or, more exactly, science theory because it is not yet grounded on facts. However, I am convinced or, better, I hope that the future can translate my theory into the enthralling language of facts.

SUMMARY OF CHAPTER IV

Symbolism is important for several reasons. It has molded and nurtured modern thought, literature, arts and sciences. It has unveiled hidden connections existing in the universe, and between arts (poetry, painting, music, opera, etc...). The collage, for instance, was inappropriately called "the surrealist collage" because it was factually an invention inherent in symbolism but bequeathed to surrealist poets and painters by symbolists. Almost every painting by Dali bears the seal of the symbolist collage. Similarly, most poems by surrealists subsume this technique. Symbolism has also inspired Picasso's cubism, Manet and Monet's impressionism. Additionally, it has proven that there is a bridge between several major areas of human knowledge and especially between cognitive sciences, literature, semiotics, and philosophy. Indeed through synesthesia, a few researchers have found and understood how the human brain functions and generates poetry when two or more of its psychic areas merge as a result of the merging/tangling of two or more synapses belonging to two or more cerebral psychic areas. They have also found the etiology and the therapy of diseases caused by a malfunction of certain areas of the brain (epilepsy and hemiplegia, etc . . .). They have come to realize that synesthesia takes place in the left hemisphere of the brain and when translated into language it becomes a very highly refined form of literary expression which is precisely poetry. Two scientists, Ramachandran and Hubbard, have worked in this direction in their paper "The Emergence of the Human Mind: Some Clues from Synesthesia". They have pertinently found that a specific region of the brain called the "angular gyrus" is very sensitive to the generation of cross-modal abstraction and cross-modal metaphors, synesthetic metaphors. Moreover, synesthesia, as a holistic way of perceiving reality, urges us to ponder how it is factually. Is it fragmented or holistic? This philosophical question was analyzed by

Plato in *The Myth of the Cave* and opens up a broad spectrum of cognitive possibilities. Finally, symbolism, through synesthesia, highlights the complexity of reality and suggests that a fruitful symbiosis of several areas of human knowledge should be necessary to decrypt it and decipher the mysteries of the universe. This can boost the advancement of knowledge in cognitive sciences, psycholinguistics, semiotics, poetry and philosophy. Consequently, symbolism is teleological and revolutionary in so far as it undergirds a new epistemological rupture and even gives researchers the opportunity to bridge science and literature through a new field of expertise called "neuroesthetics. Synesthesia has several interests: philosophical, artistic, literary, cognitive, therapeutic, scientific. It can, for instance, boost a scientific understanding of the fourth dimension, which will help solve enigmas such as teleportation, invisibility, parallel universes, NDE phenomena, time travel, creation of space-time short-cuts, tunnels (wormholes) for inter-galactic travels, etc . . .

CONCLUSION

We have undertaken an intellectual and eclectic trip revolving around the analysis of symbolism, synesthesia, under the hermeneutic guidance of semiotics. This journey has led us from literature/poetry to neuroscience, through arts, neuroesthetics, philosophy, psychology, linguistics, physics, and therapeutics. In light of this analysis, we have come to find out that:

- Symbolism is a cosmological and ambitious field of study. As a matter of fact, through synesthesia, symbolism reflects on the universe at a microcosmic and macrocosmic level to decrypt its mysteries and understand human nature and mind in particular;
- Most elements of the universe are interconnected at a microcosmic and macrocosmic level. Synesthesia symbolizes this intra-cosmic unity;
- Symbolism through synesthesia sparks a better understanding of several fields of human knowledge: neuroscience (it unravels the complexity of the human brain, contributes to the understanding the etiology of its diseases in order to decipher their therapy); physics and philosophy (the fourth dimension through space-time continuum, *monism); arts (origins of surrealism, cubism, impressionism), which clearly shows that it is at the forefront of sciences, arts and culture, poetry and language; linguistics/ semiotics (symbolism by means of synesthesia tries to generate a powerful communication by endeavoring to ward off the arbitrariness of language, motivate its signifier, and achieve its most refined stage: Cratylism);

- Synesthesia has nurtured a new epistemological break by revealing the unity and order presiding over the cosmos expressed by a deterministic perspective and the cosmic principle of interconnectedness, which implicitly helps understand the manifestations of the second law of thermodynamics (low entropy and absence of entropy). It postulates a meticulous reflection on the essence/nature of reality (holistic? fragmented? monistic?) and the *qualia as an epistemological challenge that was boldly conducted by Plato in *The Myth of the Cave;*
- If we, human beings, aspire to become the lords of the cosmos, it is necessary for us to conduct research across disciplines. One of the most tangible instances of this cross-wise intellectual enterprise can be achieved through synesthesia because it definitely calls for a fruitful collaboration between scholars from several fields: linguists, semioticians, poets, neuroscientists, medical doctors, psychologists, cosmologists, to mention but a few. Synesthesia will unquestionably ignite this collaboration on the basis of which it is purported to become a common "intellectual jurisdiction" in a convivially shared "territory".

Nowadays, even if symbolism and synesthesia in general no longer have the merit they deserve, we should not forget that they have brought an eminently significant contribution to modern literature. They have succeeded in optimizing the space that they arrogated themselves: syntax was re-arranged (especially by Mallarmé and his followers), lexicon was purified, which made it possible to achieve linguistic catharsis. This endeavor climaxed into the quasi-systematization of connotation at the expense of denotation, the coining of neologisms. Rhythm, rhyme and stanza paradigms were loosened or rejected (modernism), lexicon was purged of non-visual associations (Imagism) and canonical restraints/restrictions (Dadaism). The collage was systematized, poetic orthodoxy/classicism and rationality were rejected (Surrealism). By a sort of belated Romanticism, poetry was returned to the exploration of the realms of mysticism, *onirism, and the irrational. Even Poststructuralism was heralded by the semantic evasion from and the 'de-focusing' of the universe highlighted by the Decadent Movement and suggested by the morphology of synesthesia en style artiste. However, one of the most important elements and contributions of symbolism was the diligent work on language to achieve absolute poetry. In this respect, symbolists tried to design or standardize a linguistic arsenal. They believed that by reaching this goal, it would be possible for them to decrypt the enigmas of the universe because the mastery of the universe depends on that of language (language being an instrument, a means, and mastery an end), which was the implicit aspiration of Cratylus. In the following poem, we tried to epitomize this cryptic aspiration: Cratylus's complex

CRATYLUS'S COMPLEX

Triviality weaves our existence.
We feed on its very substance.
Language has lost its dignity
Under the diktat of paucity.
Let us distill its quintessence
And re-galvanize its essence!
We must dig the semantic abyss,
Coin new ***lexemes** from its pubis.
Lexical prostitution thwarts transcendence
And virtually necroses our existence.
For ever let us ban denotations
And abide by pure connotations!
To ward off the gap signifier-signified
Let us preclude texts from being reified!
Linguistic catharsis becomes an asset.
It must be sparked now from the onset.
Hence the urgency of a special medium
In order to win this stately ***proelium.**
Such was the goal of Master Cratylus,
His disciples, Plotinus, Catullus,
All the thaumaturges of symbolism,
And the hieratic priests of idealism.
For the sake of optimizing semantics,
Let us conceive a maiden stylistics
A new age will emerge firm and bright
If we subscribe to this semiotic right.

BONAVENTURE BALLA

Notes:

- 'Proelium' from Latin: proelium = battle, fight.
- **Lexemes = words; basic distinctive, meaningful units of a lexicon (linguistic definition).**

Such is the aspiration and historical legacy of symbolism. Nowadays, this legacy needs to be fructified within the framework of synesthesia and research conducted by experts of several fields and working in dynamic symbiosis. This powerful synergy is necessary to boost the advancement of knowledge in sciences, literature, and vicinal areas to decrypt the mysteries of the cosmos and assert our sovereignty on the universe. Consequently, symbolism is teleological and revolutionary in so far as it has very noble ideals and undergirds a new epistemological rupture. Since the symbolist poet is an inspired divine vessel, these ideals are catalyzed by a gift from heavens. Let us keep in mind that for pure symbolists (Baudelaire, Mallarmé, Rodenbach, Yeats, etc . . .), poetry, in his highest form, is nothing but a gift from heaven or at least an expression of this gift, in Latin "ex caelis oblatus" . . .

EX CAELIS OBLATUS (A GIFT FROM HEAVENS)

Exquisite Pearl fallen from heaven,
Xylophones and lyres gild your haven!

Cradle where the absolute radiates,
Auspices that no human obviates,
Emblem of elation and delectation,
Lift our parturition up to perfection!
Ignite our sheer creative enterprise
So that masterpieces always arise!

O sublime gift molded by pristine Deity,
Beacon abating our daring tribulation,
Load my lyre with divine inspiration
As it sparks the birth of this oblation,
Thurible where gold appealingly shines!
Unique work on which God lovingly signs,
Set His viaticum to eternity!

BONAVENTURE BALLA

Regarding synesthetes, in light of their fascinating faculties, we can hypothesize that they symbolize the future human species. Darwin has proven that species exponentially evolve over time and humans are not an exception. From this postulate, it can be inferred that the future human species will probably be endowed with higher faculties enabling it to live in full plenitude since the brain will certainly evolve as well. As we have mentioned in previous chapters, some researchers believe that most individuals use a minor portion of their brain/mind potential: one tenth. They cannot extend it beyond this portion (Dr. Rampa, *You, for ever*, 205). However, in light of what we know on synesthesia, it seems that synesthetes use more than this portion. Nowadays, we can scientifically assess some of their faculties in action through a portion of the spectrum of their brain capabilities: an outstanding memory (*eidetic memory, *hypermnesia -> Dr. Cytowic, Simon Baron/Cohen), the ability to see images in three dimensions, experience life in plenitude, artistically, holistically and idiosyncratically. Actually, reality is seen, heard, tasted, smelt, felt, colored simultaneously and, in some forms of synesthesia, even orgasm is experienced in color. Besides, what synesthetes feel, taste, hear, see, smell, has a mood, a gender, a "specific personality". Their brain function enhances their creativity, poetic activity (angular gyrus, frontal lobe, limbic system (amygdalae in particular)), and enables them to quickly perceive hidden objects (Dr. Ramchandran -> hidden triangles, page 121). All this is performed as if synesthetes were endowed with a sixth sense, an extra one (Drs. Ramachandran/Hubbard, Cytowic). Consequently, we believe that synesthetes might herald the future human species. It follows that on-going research will certainly turn this hypothesis into a bare fact. With progress made in this field of knowledge, we will be edified. Again, a fruitful collaboration between sciences and arts is of paramount importance to reach this goal and decrypt the mysteries of life and the universe. As a matter of fact, science explains the "how" whereas arts/philosophy explain the "why". Synesthesia is certainly one of the areas of knowledge likely to synergize the "why" and "how", science and arts. May this synergy become a fiat for our evolution!

SYNERGY

Science and arts deploy a futuristic oneness
Yawed to optimize life in its starry brightness.
Now, it behooves us to carve this golden age,
Emblem of a synergetic pilgrimage.
Rich in its quality, pregnant with density,
Guided by a divine and heuristic audacity,
Yeast of future, grow the seeds of felicity!

BONAVENTURE BALLA

ILLUSTRATIONS OF SYMBOLISM WITH SYMBOLIST TEXTS:

I—*CORRESPONDANCES* by Baudelaire
La nature est un temple où de vivants piliers
Laissent parfois sortir de confuses paroles
L'homme y passe à travers des forêts de symboles
Qui l'observent avec des regards familiers.

Comme de longs échos qui de loin se confondent
Dans une ténébreuse et profonde unité,
Vaste comme une nuit et comme la clarté,
Les parfums, les couleurs et les sons se répondent.

Il est des parfums frais comme de chairs d'enfants,
Doux comme les hautbois, verts comme les prairies,
—Et d'autres, corrompus, riches et triomphants,

Ayant l'expansion des choses infinies,
Comme l'ambre, le musc, le benjoin et l'encens,
Qui chantent les transports de l'esprit et des sens.

CORRESPONDENCES

Nature is a temple where living pillars
Let escape sometimes confused words;
Man traverses it through forests of symbols
That observe him with familiar glances.

Like long echoes that intermingle from afar
In a dark and profound unity,
Vast like the night and like the light,
The perfumes, the colors and the sounds respond.

There are perfumes fresh like the skin of infants
Sweet like oboes, green like prairies,
—And others corrupted, rich and triumphant

That have the expanse of infinite things,
Like ambergris, musk, balsam and incense,
Which sing the ecstasies of the mind and senses.

Note: The Sweden philosopher and mystic Immanuel Swedenborg revealed the notion of Correspondences to his disciples and followers.

EXPLORATION OF THE TEXT: 2 levels: hermeneutic and heuristic

1—HERMENENEUTIC READING:

At the hermeneutic level, we read the passage for the first time and try to understand its meaning. At this stage we come up with a number of interpretations. The following questions are intended to spark these interpretations:

*<u>QUESTIONS</u> -> *Answers to all questions can be found at the end of the book

    A—What are the nature, main idea, and structure of this poem?
    B—This poem epitomizes symbolism in terms of its theoretical program and esthetic canons. On the basis of information from this text, point out 5 major canons of symbolism and summarize them.
    C—The very first word of the text is "Nature"? What is the semantic implication related to the use of this word? What is the stylistic device used in the first two lines of the first stanza?

2—HEURISTIC READING

At the heuristic level, we do a second or even a third reading of the text, a retroactive one, which makes us to move to a deeper level and decrypt other layers of meanings. At this stage we understand the text much better because we 'discover' or 'find out' what the text really is (from the Greek "eureka" meaning "I found out", "I discovered" by analyzing it, confronting its ideas, concepts, and notions with ideas, concepts, and notions that we have acquired from other texts. Then, such a confrontation enables to find possible connections with other texts (intertextuality), which leads to interesting and many deeper findings in terms of meaning (s) and understanding. The following questions will help to trigger these findings.

QUESTIONS

a—The construction "Nature is a temple" develops the concept of semantic indirection. As you know, semantic indirection creates a distension within the metaphorical discourse. How can you analyze this concept in light of what you have studied regarding isotopies? Into how many isotopies can you break down this construction? Identify and examine them. How can you measure the density of poetic shocks thus created? Specify/appreciate the effect of poetic shocks.

b—What is the metaphysical implication of the expression: "Nature is a temple"?

c—What do the constructions "confuses paroles", "forêts de symboles" suggest?

d—To what extent is the word 'unity' important in the second stanza? What does the 4th line of this stanza stand for? Which word nurtures the *semantic matrix of the 4th line? How?

e—The title of the poem is 'Correspondences'. How does it help to deepen the understanding of the text. How many types of correspondences are there in the text?
Identify them. In what stanzas do they occur?

f—The apex and aspiration of symbolism is Cratylism, that is, the attempt to reduce the arbitrariness of the connection between signifier and the signified and, by the same token, to maximize the power of language. How is Cratylism manifested in this poem?

II—*L'ALBATROS* by **BAUDELAIRE**

Souvent pour s'amuser les hommes d'équipage
Prennent des albatros, vastes oiseaux des mers
ui suivent, indolents compagnons de voyage,
Le navire glissant sur des gouffres amers.

A Peine les ont-ils déposé sur les planches
Que ces rois de l'azur, maladroits et honteux
Laissent piteusement leurs grandes ailes blanches
Comme des avirons trainer à côté d'eux.

Ce voyageur ailé, comme il est gauche et veule!
Lui naguère si beau, qu'il est comique et laid!
L'un agace son bec avec un brûle-gueule,
L'autre mime en boîtant l'infirme qui volait!

Le Poète est semblable au prince des nuées
Qui hante la tempête et se rit de l'archer,
Exilé sur le sol au milieu des huées,
Ses ailes de géant l'empêchent de marcher.

EXPLORATION OF THE TEXT: 2 levels (See answers at the end of the book)

1—HERMENENEUTIC READING:

At the hermeneutic level, we read the passage for the first time and try to understand its meaning. At this stage we come up with a number of interpretations. The following questions are intended to spark these interpretations:

QUESTIONS:

a—What are the nature, structure and main idea of this poem?
b—What does the periphrastic construction 'hommes d'equipage' stand for? What is the meaning of 'prennent'. Define the word 'albatross.
c—This poem epitomizes symbolism in terms of its theoretical program and esthetic canons. On the basis of your knowledge of symbolism, can you summarize, in 4 major canons, the way it is presented in this text ? How?

2—HEURISTIC READING

At the heuristic level, we do a second or even a third reading of the text, a retroactive one, which makes us to move to a deeper level and decrypt other layers of meanings. These questions will try to trigger deeper findings.

QUESTIONS

a—Identify and examine synesthesia in the first line. How can you measure the density of poetic shocks thus created? Specify/appreciate the effect of poetic shocks.
b—Symbolism: What do these words stand for: "hommes d'equipage", "planches", "rois de l'azur", "prince des nuees"? The construction "infirme qui volait" introduces a contradiction. Explain this contradiction. Explain the construction "ses ailes de geant l'empechent de marcher".

c—Is this poem allegorical or symbolical? Justify your answers in light of your knowledge of symbols and allegory. Explain its layers of meaning on the basis of the meaning of the albatross and that of "les planches"? As you know the aspiration of symbolism as a poetic school is Cratylism. How is cratylism manifested in this poem?

Study synesthesiametrics in light of three major canons/benchmarks at least.

III—*LA VIE ANTERIEURE* by Baudelaire

J'ai souvent habité sous de vastes portiques
Que les soleils marins teignaient de mille feux,
Et que leurs grands piliers, droits et majestueux,
Rendaient pareils le soir, aux grottes basaltiques.

Les houles en roulant les images des cieux
Mêlaient d'une façon solennelle et mystique,
Les tout-puissants accords de leur riche musique
Aux couleurs du couchant réflété par mes yeux[. . .]

C'est là que j'ai vécu dans les voluptés calmes,
Au milieu de l'azur, des vagues, des splendeurs
Et des esclaves nus, tout imprégnés d'odeurs,

Qui me rafraîchissaient le front avec des palmes,
Et dont l'unique souci était d'appronfondir
Le secret douloureux qui me faisait languir

EXPLORATION OF THE TEXT:

1—HERMENEUTIC READING

QUESTIONS:

a—What are the nature, structure and main idea of this poem?
b—This poem epitomizes symbolism in terms of its theoretical program and esthetic canons. On the basis of your knowledge of symbolism, can you summarize, in 4 major canons, the way it is presented in this text ? How? (suggestions: idealism; music; phonic mimologism, synesthesia -> Cratylism; sophisticated vocabulary; free verses vs. freed verses => psychological rhythm vs. metrical rhythm)
c—The first two lines of the first quatrain refers to the setting. What information does the text provide regarding the characteristics of this setting? Examine lexical items and their semantic field to find this information. What is the stylistic device used in the second stanza? What does it suggest?

2—HEURISTIC READING

QUESTIONS

a—The title "Vie antérieure" (Previous life) suggests a seminal concept and notion inherent in idealism and Platonism and symbolism. Explain it.
b—Which words or constructions nurture the semantic matrix leading to the generation of the whole poem?
c—The poem maximizes the use of music. What are the prosodic and literary implications of this process
d—What is the symbolism of the construction "esclaves nus"

APPLICATION OF SYNESTHESIAMETRICS

Apply synesthesiametrics to the poem with three major esthetic canons

IV—CHANT D'AUTOMNE by BAUDELAIRE

Bientôt nous <u>plongerons dans les froides ténèbres</u>;
Adieu, vive clarté de nos étés trop courts!
J'entends déjà tomber avec des chocs funèbres,
Le bois retentissant sur le pavé des cours,

Tout l'hiver va rentrer dans mon être: colère,
Haine, frissons, horreur, labeur dur et forcé,
Et comme le soleil dans son enfer polaire,
Mon coeur ne sera plus qu'un bloc rouge et glacé.

J'écoute en frémissant chaque bûche qui tombe;
L'échafaud qu'on bâtit n'a pas d'écho plus sourd
Mon esprit est pareil à la tour qui succombe
Sous les coups d'un bélier infatigable et lourd.

Il me semble, bercé par ce choc monotone,
Qu'on cloue en grande hâte un cercueil quelque part...
Pour qui?. C'étaiat hier l'été, voici l'automne!
Ce bruit mysterieux sonne comme un départ

EXPLORATION OF THE TEXT:

1—HERMENEUTIC READING

a—What is the nature of the poem? What is its main idea? Justify the title. Explore the *semantic field of grief. Identify the stylistic device contained in the first line. Identify its components. What does it suggest?

b—This poem epitomizes symbolism in terms of its theoretical program and esthetic canons. On the basis of your knowledge of symbolism, can you summarize, in 4 major canons, the way it is presented in this text ? How?

c—First stanza: line #1: Study the tense used in the first stanza in relation to the rest of the stanza and all the other stanzas

2. What does this tense suggest in terms of meaning and tone? Identify the stylistic device contained in the first line. Identify its components. What does it suggest?

2—HEURISTIC READING

Apply synesthesiametrics to the passage.

V—SONNET DES VOYELLES by ARTHUR RIMBAUD

A noir, E blanc, I rouge, U vert, O bleu: voyelles,
Je dirai quelque jour vos naissances latentes:
A, noir corset velu des mouches éclatantes
Qui bombinent autour des puanteurs cruelles,

Golfes d'ombre; E, candeur des vapeurs et des tentes,
Lances des glaciers fiers, rois blancs, frissons d'ombelles;
I, pourpres, sang craché, rire des lèvres belles
Dans la colère ou les ivresses pénitentes;

U, cycles, vibrements divins des mers virides,
Paix des pâtis semés d'animaux, paix des rides
Que l'alchimie imprime aux grands fronts studieux;

O, suprême Clairon plein des strideurs étranges,
Silence traversés des Mondes et des Anges:
—O l'Oméga, rayon violet de Ses Yeux !

A. Rimbaud

EXPLORATION OF THE TEXT:

I—HERMENEUTIC READING

    a—Specify the nature, structure and main idea of the text.
    b—Summarize three or four esthetic canons of symbolism in this poem.

II—HEURISTIC READING

Study the poetic quality, the semantic density, and idiolectic pertinence of constructions in this sonnet.

How does this poem attempt to achieve Cratylism? What is the poetic quality of the text?

Show the idiolectic pertinence of the passage.

VI—*THE SECOND COMING* by William Butler Yeats (1865-1939)

TURNING and turning in the widening gyre
The falcon cannot hear the falconer;
Things fall apart; the centre cannot hold;
Mere anarchy is loosed upon the world,
The blood-dimmed tide is loosed, and everywhere
The ceremony of innocence is drowned;
The best lack all conviction, while the worst
Are full of passionate intensity.

Surely some revelation is at hand;
Surely the Second Coming is at hand.
The Second Coming! Hardly are those words out
When a vast image out of *Spiritus Mundi*
Troubles my sight: somewhere in the sands of the desert
A shape with lion body and the head of a man,
A gaze blank and pitiless as the sun,
Is moving its slow thighs, while all about it
Reel shadows of the indignant desert birds.
The darkness drops again; but now I know
That twenty centuries of stony sleep
Were vexed to nightmare by a rocking cradle,
And what rough beast, its hour come round at last,
Slouches towards Bethlehem to be born?

EXPLORATION OF THE TEXT

I—HERMENEUTIC READING

1—what is the nature of the poem? What is its main idea? How about its structure?
2—What does the title suggest?
3—Provide a few esthetic canons of symbolism

II—HEURISTIC READING

1—Decipher the semantic nucleus/matrix of the passage. The first stanza is a powerful expression of chaos. How is this chaos suggested stylistically?
2—What is the meaning of spiritus mundi? Study its deeper meaning. How is this meaning related to the symbolist conception of the poet?
3—There is a seminal isotopy in the passage. Identify it. What words or constructions endorse this isotopy?
4—Specify the seminal isotopy subsumed in the poem. What words and constructions endorse it?
5—To what extent can this passage contribute to modern literature?
6—How is Cratylism manifested in the poem?

VII—*CIMETIERE MARIN* by PAUL VALERY

Ce ciel tranquille, où marchent des colombes,
Entre les pins palpite, entre les tombes;
Midi le juste y compose de feux
La mer, la mer, toujours recommencée!
O récompense après une pensée
Qu'un long regard sur le calme des dieux!

EXPLORATION OF THE TEXT

I—HERMENEUTIC READING

- Provide the nature of the poem, its main idea and structure.
- Since this poem expresses the notion of peace and vitality, what lexical items suggest these notions?
- Outline three (3) or four (4) esthetic canons of symbolism in this passage

II—HEURISTIC READING

-What is the semantic nucleus of the poem? Note that the semantic nucleus or matrix is the focal point from which the meaning of the whole text stems.
-Point out the isotopy that nurtures this semantic nucleus.
-Study *intertextuality through this passage.

VIII—*BRISE MARINE* by Stephane Mallarmé

#La chair est triste, hélas! et j'ai lu tous les livres.
Fuir! là-bas fuir! Je sens que des oiseaux sont ivres
D'être parmi l'écume inconnue et les cieux!
Rien, ni les vieux jardins reflétés par les yeux
Ne retiendra ce coeur qui dans la mer se trempe.
O nuits! ni la clarté déserte de ma lampe
Sur le vide papier que la blancheur défend
Et ni la jeune femme allaitant son enfant.
Je partirai! Steamer balancant ta mature,
Leve l'ancre pour une exotique nature!
Un Ennui, désolé par les cruels espoirs,
Croit encore à l'adieu suprême des mouchoirs!
Et, peut-être, les mâts, invitant les orages,
Sont-ils de ceux qu'un vent penche sur les naufrages.
Perdus, sans mâts, sans mâts, ni fertile îlots.
Mais, O mon coeur, entends le chant des matelots!

EXPLORATION OF THE TEXT:

I—HERMENEUTIC READING

1—Specify the nature of the passage, its structure and main idea.
2—What are the constructions and words used to stand for this ideal world?
3—Provide three major esthetic canons of symbolism of this passage.

II—HEURISTIC READING

1—Intertextuality. This passage has connections with a few poems written by Baudelaire. Specify these texts and the nature of these connections.
2—How is Cratylism achieved in the poem?

IX—Excerpt from *BRUGES-LA-MORTE* by GEORGES RODENBACH

Une flamme lui chanta aux Oreilles.
Un picotement lui vint aux yeux.
Il sentit un brouillard contagieux lui entrer dans l'#âme;
envahi par le silence froid.

EXPLORATION OF THE TEXT

I—HERMENEUTIC READING

- What is the nature of the passage?
- Specify the structure and main idea of the poem.

II—HEURISTIC READING

- This short text is composed of synesthesia. How is it structurally presented? Analyze the specificity of this synesthesia. In light of what you know about the Decadent Movement, conduct a diligent study of this short passage by enriching or extending it with additional information from the esthetics of this movement.
- What is the contribution of this text to modern literature and modernity in general?

GLOSSARY

Adynaton: (plural 'adynata') term stemming from the Greek 'adynaton' meaning 'impossible'. Stylistic device whereby a narrator depicts a situation, an event factually impossible to happen or enshrouded by the mystery or the supernatural. e.g. **"the wolf is escaping from lambs"**

Alliterations vs. assonances: combination of words beginning with consonantic sounds (or predominantly made of consonants) within a sentence or verse as opposed to assonances (involving the combination of vocalic sounds) to suggest a special effect.

Amygdalae: the amygdalae are two small, round structures located anterior to the hippocampi near the temporal poles. The amygdalae are involved in detecting and learning what parts of our surroundings are important and have emotional significance. They are critical for the production of emotion, and may be particularly so for negative emotions, especially fear.

Anacoluthon (plural 'anacolutha'): Sudden change in the construction of a sentence for the sake of emphasis. **"Exiled in the ground [....]/ His gigantic wings prevent him from soaring"**

Angular gyrus: a ridge in the neocortex in the parietal lobe, next to the temporal and occipital lobe, concerned with the position of the body in space and **linking** sound and meaning. This linking is very important because it suggests its cross-modal function in metaphors and cross-modal abstraction

Anomia: From the Greek 'a/an' = 'absence, suppresion' and 'nomos' = 'name'. Disease characterized by the loss of the faculty enabling an individual to name objects. The patient still knows and recognizes objects whenever he/she sees them but he/she just cannot give them a name. This can happen if the angular gyrus is damaged.

Anterior cingulate—The anterior cingulate cortex (ACC) is located in the middle of the brain, just behind the prefrontal cortex. The ACC is thought to play a central role in attention, and may be particularly important with regard to conscious, subjective emotional awareness. This region of the brain may also play an important role in the initiation of motivated behavior

Aphoristic: Conceived to express a general truth. Proverbs have an aphoristic function.

Asymbolia: Very poor and narrow conception/reading of literature: impossibility for a work to have more than one meaning, which leads to absence of symbolism or any kind of symbolic meaning. The Greek prefix "a/an" means absence, suppression, lack of, beside, aside => absence of symbol or symbolism

Asyndeton (plural 'asyndeta'): suppression of all particles or any construction likely to conjoin words or ideas in a verse or a sentence (and, or, but, neither . . . nor, either . . . or, for, etc . . .) to suggest the idea of separation.

Brainwave entrainment: brainwave entrainment or "brainwave synchronization," is any practice that aims to cause brainwave frequencies to fall into step with a periodic stimulus having a frequency corresponding to the intended brain-state (for example, to induce sleep), usually attempted with the use of specialized software. It purportedly depends upon a "frequency following" response on the assumption that the human brain has a tendency to change its dominant EEG frequency towards the frequency of a dominant external stimulus.

Calligram: remarkable osmosis between the signifier and signified, the form and the substance. The theme of the text was spatially and pictographically represented by a special array of words and lines of the passage. Thus, a poem became a privileged space where both the canvas and a graphic system were unified harmoniously and artistically.

Cathartic function: According to Aristotle, theater was designed to purge us (spectators) of our passions, to purify the soul through intense emotional feelings and, by the same token, teach and heal us. Therefore, he assigned it a didactic and therapeutic function known as "cathartic". The etymological meaning of the word "catharsis" is very suggestive. In Greek "catharsis" means "quietness of the soul". Factually, a soul purified of its passions is quiet, calm.

Cerebellum—Recently, there has been a considerable amount of work that describes the role of the cerebellum in emotion as well as cognition, and a "Cerebellar Cognitive Affective Syndrome" has been described. Both neuroimaging studies as well as studies following pathological lesions in the cerebellum (such as a stroke) demonstrate that the cerebellum has a significant role in emotional regulation. Lesion studies have shown that cerebellar dysfunction can attenuate the experience of positive emotions. While these same studies do not show an attenuated response to frightening stimuli, the stimuli did not recruit structures that normally would be activated (such as the amygdalae). Rather, alternative limbic structures were activated, such as the ventromedial prefrontal cortex, the anterior cingulate gyrus, and the insula. This may indicate that evolutionary pressure resulted in the development of the cerebellum as a redundant fear-mediating circuit to enhance survival. It may also indicate a regulatory role for the cerebellum in the neural response to rewarding stimuli, such as money, drugs of abuse, and orgasm.

Concatenation: a series of interconnected things or events (from the Latin 'catena' = chain); logic connection or sequel of ideas manifesting a metaphysical kinship; stylistic device whereby repetition is systematized in such a way that it creates a complete circle, a loop or chain (Latin catena)

Corticospinal tract: pathway composed of the spinal cord and the cortex

Creative idiosyncrasy: one's own way of expressing one's creativity

Deictic particle: a small word conceived to point to, beckon, show, or designate something in order to focus the attention of listener on it e. g. "ci ", "là" in French, "**Aquí, allí**" in Spanish, **"here"**, **"there"** in English, **"Hic"**, **"ibi"** in Latin. Deictic particles can function as adverbs.

Divisionism: trend within impressionism whereby the separation of colors through strokes of pigments are used to create a scene of balance and statis.

Dyslexia: disease characterized by the fact that the subject suffers from dysfunctional reading or trouble in reading conveniently

Eidetic memory: From the Greek "eidos" = "essence". Exceptional memory endowing the subject with the ability to remember everything he has seen or heard before with every single minute detail and slight difference, which precisely fits the synesthetes.

Epistemological from 'epistemology': theory of knowledge, the critique of knowledge, section of philosophy dealing with the critique of knowledge and sciences. (from Greek ἐπιστήμη *(epistēmē)*, meaning "knowledge, understanding", and λόγος *(logos)*, meaning "study", "speech", "word") is the branch of philosophy dealing with the nature and scope (limitations) of knowledge. It essentially focuses on: the foundations of knowledge, its presuppositions and basis, and the general reliability of claims of knowledge. It is one of the five (5) classical fields of philosophical inquiry (**epistemology**, esthetics, ethics, logic, and metaphysics). It tries to provide answers regarding:

- The nature of truth;
- The essence of knowledge;
- How it is acquired;
- How it can be verified;
- What are its limits;
- The relationships between the knower and the known. In epistemology knowledge is supposed to be conceived as the knowledge of what is true, but the knowledge of truth itself is for the most part regarded as more the jurisdiction of logic and metaphysics than epistemology.

Eutaxy: From the Greek 'eu' = 'beautiful' and 'taxis' = 'arrangement'. Perfect harmony governing a system, each element being perfectly symmetrical to others. Such harmony presides over each element taken individually and the whole system. In the final analysis, high order prevails internally and externally. Eutaxy is suggested by the Pythagorean harmony and the reference to architectonics.

Euchronic or U-chronic vision: From the Greek 'eu' = 'beautiful', 'u' = 'absence/ outside' and 'chronos' = time/ history. Subjective perspective whereby the subject perceives reality and history not the way it is but the way he wished them to be. It is a kind of false consciousness. The connotation of this word is close to the pejorative conception of ideology in Marx's texts. As a matter of fact, for Marx and Friedrich Engels, ideology is a false consciousness, a construct, a fiction because it is a system that purports to accurately reflect reality but distorts it, creating a false consciousness.

Hemiplegia: brain disease causing a paralysis of one hemisphere resulting in the paralysis of the opposite side of the body. If the left hemisphere is paralyzed then the right side of the body is paralyzed.

Hermeneutic reading vs. heuristic reading. Hermeneutic reading can be equated with a first reading. I call it proto-reading (From the Greek 'protos' = first). It gives the reader, cryptanalyst the opportunity to get acquainted with it and start interpreting it. At this stage, the reader can understand it but generally (not always) in a superficial fashion.

Heuristic (from the Greek "eureka" meaning "I found out", "I discovered) **reading:** a deeper reading of the text, a retroactive reading. Generally, it is a second or third one. At this stage, the reader compares and contrasts it with other texts (intertextuality), historicizes (sociohistorical context), analyzes its specific signs to decrypt the meaning and how it functions for the purpose of communication (stylistics and semiotics). Finally, he/she manages to understand its meaning and layers of meaning. Heuristic reading entails a much deeper quest for meaning (s) by using a whole linguistic arsenal. It is specially suitable for poetry. Since understanding a poetic text is not a matter of course, the meaning can be deceptive if it is not meticulously analyzed. It follows that a poet usually recurs to semantic indirection: he says something to mean something else.

Homometrical structure: use of the same metrical structure throughout a text (binary, ternary rhythm, etc . . .)

Homonymic or paronymic collision: use of words having a graphological kinship by brutally putting one next to another in a verse or sentence (clore/clouer) Palindrome: poem, verse or just a sentence in which one uses words that can be read from left to right indistinctively (e.g. **Laval, rotor**) to produce a specific effect, usually a musical or ludic one.

Homophonic phrases: use of words or constructions having the same sounds but not necessarily the same meaning or spelling (**mettre/ metre [metr]** in French right/wright **[rait]** in English)

Hyperbaton (plural 'hyperbata'): inversion of a sentence/verse structure to put a special emphasis on a notion, concept, or idea e.g. "This I will do "instead of "I will do this"

Hyper-literalism: see polysemy and monosemy

Hypermnesia: exceptionally developed memory usually under abnormal conditions (trauma, hypnosis)

Hypnagogic experience: experience in which the subject is in an intermediate state, between wake and sleep. It is a subliminal state.

Ideophone: a sound that suggests an idea.

Imitative harmony: device through which a poet uses recurrent patterns to imitate specific sounds or spellings

Immanentism: Philosophical system positing that God is inside nature. Immanentism is also a form of pantheism, Spinoza' s philosophical doctrine. Pantheism posits that God is inside nature and everywhere in nature (in Greek "pan" means "everywhere, all"). Spnioza used to summarize it with his Latin motto "Deus sive natura", that is: "God or nature".

Insula—The insular cortex is thought to play a critical role in the bodily experience of emotion, as it is connected to other brain structures that regulate the body's autonomic functions (heart rate, breathing, digestion, etc.). This region also processes taste information and is thought to play an important role in experiencing the emotion of disgust.

Isotopy: The notion of *isotopy refers to the repetition of a basic meaning trait (seme); such repetition, establishing some level of familiarity within a text, allows for a uniform reading/interpretation of it. An example of a sentence containing an isotopy is *I drink some water*. The two words *drink* and *water* share a seme (unit of meaning and a reference to liquids), and this gives homogeneity to the sentence. This concept, introduced by Greimas in 1966, had a major impact on the field of semiotics, and was redefined multiple times. Catherine Kerbrat-Orecchioni extended the concept to denote the repetition of not only semes, but also other semiotic units (like phonemes for isotopies as rhymes, rhythm for prosody, etc.). Umberto Eco showed the flaws of using the concept of "repetition", and replaced it with the concept of "direction", redefining isotopy as "the direction taken by an interpretation of the text".

Lexeme: distinctive and meaningful basic unit of the lexicon (specialized vocabulary). A lexeme is a word belonging to a specific vocabulary ot just a word in general.

Lexical vicinity: the proximity of a word with other words. It can help in elucidating its meaning by contextualizing it and, by the same token, using its semantic field.

Lexico-semantic: refers to lexemes (basic distinctive meaningful units of a lexicon) and their meaning enlightened by a specific context (semantics). So, essentially, it deals with words and their meaning. Words are surrounded by a field of meaning nurtured and enlightened by a specific context.

Minimal pair: also known as "pair of quasi-homonyms" are words having almost the same spelling except for one letter **e.g. "sake"** and **"take".**

Motu proprio: Latin phrase meaning: "out of/from its own movement" and, by extension, by itself.

Natura naturans as opposed to natura naturata. Natura naturans: perpetually changing, evolving, and open-ended nature whereas **"natura naturata"** designates an accomplished nature, a nature that has completed its final stage and cannot, by the same token, evolve anymore. That is why it has a **"ne varietur"** essence, which means cannot be changed.

Nomenon: In Kant's philosophy, this concept refers to the unknown, what the human mind/reason cannot access, subsumed in 3 notions:

- The nature of the soul,
- God and,
- The universe.

Onirism: systematic occurrence of dreams.
Ophicleides: obsolete bass instrument of music with keys
Oxytocin: a neurotransmitter involved in social bonding.
Palingenesis: From the Greek "palin" = "again" and "genesis" origin. Theory that posits that history is nothing but a succession of endless cycles, co-cycles, and repetitions.
"Paronymic derivative": the expression "paronymic derivative" stems from Aristotle's teaching. Indeed, Aristotle considered that words can be created on the basis of their roots, and this process helps students to understand and spell all words similar in terms of spelling. For instance words like 'philosophical' 'philosophically' stems from 'philosophy'. It follows that if a student knows how to spell philosophy, he will certainly know how to spell philosophical and philosophically as well because 'philosophical' and 'philosophically' have the same stem: 'philosophy'. Therefore

'philosophical' and 'philosophically' were created by paronymic derivation. Likewise 'symbolism', 'symbolization', 'symbolical', are paronymic derivatives of 'symbol'

Phonic mimologism is conceived to particularly stress phonic similarities linked with paronymic derivation, a technique formalized by Aristotle during the Greek Antiquity to create words on the basis of paronyms: words morphologically akin in terms of their etymology. It systematically assembles words by taking into account their musical kinship. Through this process, the poet elicits amazing effects from the suggestive power of sounds.

Pineal gland: A pea-sized gland located near the thalamus that produces melatonin, which regulates the sleep-wake cycle.

Pituitary gland: hypothalamus nucleus that produces hormones, including oxytocin.

Pointillism: trend within impressionism, in which a picture built up from small dots of paint that resolve into recognizable forms when viewed as a whole.

Palingenetic (see palingenesis)

Pantheism: a philosophical system positing that God is everywhere, in nature and operates within the world He has created.

Paradigmatic axis/relationships vs. syntagmatic axis: human language is organized according to two axes: paradigmatic axis and syntagmatic axis. The paradigmatic axis is composed of interchangeable elements in a vertical perspective (axis of substitution) whereas the syntagmatic axis involves elements that follow one another in the horizontal, linear perspective (axis of succession).

Paratactic structure (see parataxis)

Parataxis: (from the Greek "para" = "aside, beside" and "taxis" = "arrangement") arrangement of sentences, phrases or words, by positioning them side by side to create perfect parallelism or perfect parallel structures. One of the most famous instances of a parataxis in English is Caesar's formula: "veni, vidi, vici" = "I came, I saw, I conquered".

Paronymic derivative: word having the same morphological kinship as another word e.g. famous -> fame -> from the Latin "fama": "rumor", "noise", "news"; eat -> edible -> from "edere" in Latin: "to eat"; election -> eligible -> elect-> from the Latin "elegere"/"electum"(supine): to choose.

Paronomasia: stylistic device in which one brings words having similar sounds together.
Phonic mimologism: systematic use of similar sounds or words/ constructions having phonetic kinships, or iterative techniques to imitate reality and try to reduce the gap between the signifier and the signified (rhyme scheme, imitative harmony, alliterations, assonances, paronomasia, onomatopoeia, ideophones, palindromes, isometrical rhymes, synesthesia, etc . . .).
Pictographic technique: a technique whereby an art integrates the insertion of picture, images, or drawing.
Polysemy: one of the most important assets of literature: capability for a text to have many meanings or layers of meaning. The opposites are monosemy and **asymbolia** inherent in bad literature, bad texts. Asymbolia is the corollary of hyper-literalism. Literature distances itself from **hyper-literalism**, which is a narrow and poor reading a text and, on the contrary eminently endorses **polysemy.**
Prefrontal cortex—The term prefrontal cortex refers to the very front of the brain, behind the forehead and above the eyes. It appears to play a critical role in the regulation of emotion and behavior by anticipating the consequences of our actions. The prefrontal cortex may play an important role in delayed gratification by maintaining emotions over time and organizing behavior toward specific goals.
Qualia: singular "**quale**", from a Latin word meaning for "what sort" or "what kind," is a term used in philosophy to refer to subjective conscious experiences as 'raw feels'. Daniel Dennett writes that *qualia* is "an unfamiliar term for something that could not be more familiar to each us: the ways things seem to us."
Root Mean Square: in mathematics, the **root mean square** (abbreviated **RMS** or **rms**), also known as the **quadratic mean**, is a statistical measure of the magnitude of a varying quantity. It is especially useful when variates are positive and negative, e.g., sinusoids. RMS is used in various fields, including electrical engineering.
Semantic field: field of meaning. Each word is surrounded by a field of meaning determined by the context. The context is designed to clarify, unravel the meaning of a word or words.
Semantic indirection: semiotic principle by virtue of which one says something but means something else. The meaning usually shifts from the literal to the figurative.

Serotonin: Serotonin or **5—hydroxytryptamine (5—HT)** is a monoamine neurotransmitter. Biochemically derived from tryptophan, serotonin is primarily found in the gastrointestinal (GI) tract, platelets, and in the central nervous system (CNS) of animals including humans. It is popularly thought to be a contributor to feelings of well-being and happiness.[5]

Approximately 90% of the human body's total serotonin is located in the enterochromaffin cells in the gut (alimentary canal), where it is used to regulate intestinal movements.[6][7] The remainder is synthesized in serotonergic neurons of the CNS, where it has various functions. These include the regulation of mood, appetite, and sleep. Serotonin also has some cognitive functions, including memory and learning. Modulation of serotonin at synapses is thought to be a major action of several classes of pharmacological antidepressants.

Serotonin secreted from the enterochromaffin cells eventually finds its way out of tissues into the blood. There, it is actively taken up by blood platelets, which store it. When the platelets bind to a clot, they disgorge serotonin, where it serves as a vasoconstrictor and helps to regulate hemostasis and blood clotting. Serotonin also is a growth factor for some types of cells, which may give it a role in wound healing.

Signifier vs. signified: in linguistics and semiotics, the signifier is the expression, the form, or the letter whereas the signified is the substance, the meaning or the spirit.

Syntagmatic axis/relationships (see paradigmatic axis)

Ventral striatum—The ventral striatum is a group of subcortical structures thought to play an important role in emotion and behavior. One part of the ventral striatum called the nucleus accumbens is thought to be involved in the experience of goal-directed positive emotion. Individuals with addictions experience increased activity in this area when they encounter the object of their addiction.

ANSWERS TO QUESTIONS ON SYMBOLIST TEXTS:

I—*"CORRESPONDANCES"* by Baudelaire
La nature est un temple où de vivants piliers
Laissent parfois sortir de confuses paroles
L'homme y passe à travers des forêts de symboles
Qui l'observent avec des regards familiers.

Comme de longs échos qui de loin se confondent
Dans une ténébreuse et profonde unité,
Vaste comme une nuit et comme la clarté,
Les parfums, les couleurs et les sons se répondent.

Il est des parfums frais comme de chairs d'enfants,
Doux comme les hautbois, verts comme les prairies,
—Et d'autres, corrompus, riches et triomphants,

Ayant l'expansion des choses infinies,
Comme l'ambre, le musc, le benjoin et l'encens,
Qui chantent les transports de l'esprit et des sens.

EXPLORATION OF THE TEXT: 2 levels (See answers at the end of the book): hermeneutic and heuristic

1—HERMENENEUTIC READING:

At the hermeneutic level, we read the passage for the first time and tries to understand its meaning. At this stage we come up with a number of interpretations. The following questions are intended to spark these interpretations

QUESTIONS -> Answers are written in blue <u>Nota</u>: these answers are <u>not</u> mathematical truths. However, albeit not absolute, they are cogently supported by a meticulous semiotic analysis, broad knowledge, and logic.

    A—What are the nature, structure and main idea of this poem?

        Nature of the passage: Sonnet, that is a poem inherited from Italy and formalized during the Renaissance, composed of 2 quatrains and 2 tercets, its rhyme scheme is (abba) (cddc) (efe) (ghh)

        Main idea: perfect harmony, order and unity existing in nature.

        Structure: 3 semantic units:

Quatrain # 1:	description of nature -> this description highlights its sacredness.
Quatrain # 2:	presentation of horizontal correspondences also known as synesthesia
Tercets # 3 & 4:	presentation of vertical correspondences, those between the material and the spiritual

    B—This poem epitomizes symbolism in terms of its theoretical program and esthetic canons. On the basis of information from this text, can you summarize symbolism in 5 major canons? How?

1—Revelation of relationships between the material and the ideal world,
2—The use of synesthesia,
3—The power of suggestion instead of description or narration through the use of music: neat rhyme scheme; alliterations (parfums frais . . . chairs d'enfants); paratactic units:

> a—*Les parfums, les couleurs et les sons se répondent.*
> b—*Doux comme les hautbois, verts comme les prairies,*
> c—*Et d'autres, corrompus, riches et triomphants,*
> d—*Comme l'ambre, le musc, le benjoin et l'encens,* the corollary of all these musical elements (rhyme scheme, alliterations, paratactic units gives way to:
> e—*eurhythmy (coherent organization of musical elements conceived to generate a beautiful sense of rhythm) and *euphony (overall beauty of sounds constructed by their coherent organization)

4—The use of symbols: two instances of symbol: nature and synesthesia. Nature is a metaphor. Its highly refined form in the text grants it the status of a symbol: that of pre-established harmony, divine presence, perfect unity and organization existing in the universe. Nature also symbolizes the universe itself. Synesthesia symbolizes the correspondences between the material and the supra-terrestrial as well as intra-cosmic unity,
5—The use of suggestive lyricism is stylistically expressed by an extended metaphor, synesthesia, paratactic constructions, eurhythmy, euphony, etc . . .

C—The very first word of the text is "nature"? What is the semantic implication related to the use of this word? What is the stylistic device used in the first two lines of the first stanza?

Nature is perceived as a temple, a sacred place where men commune with God, which implies that it is housed by God, that is, perfect beauty. Therefore, it bears God's seal and, by the same token, that of perfection itself. The stylistic device supporting this idea is an *extended metaphor: "La nature est un temple ou de vivants piliers/Laissent parfois sortir de confuses paroles".

2—HEURISTIC READING

At the heuristic level, we do a second or even a third reading of the text, a retroactive one, which makes us to move to a deeper level and decrypt other layers of meanings. At this stage we understand the text much better because we 'rediscover' the text by analyzing it, confronting its ideas, concepts, and notions with ideas, concepts, and notions that we have acquired from other texts. Then, such a confrontation enables to find possible connections with other texts (intertextuality), which leads to interesting and much deeper findings in terms of meaning (s) and understanding. The following questions will help to trigger these findings.

QUESTIONS

A—The construction "Nature is a temple" develops the concept of semantic indirection. As you know, semantic indirection creates a distension within the metaphorical discourse. How can you analyze this concept in light of what you have studied regarding isotopies? Into how many isotopies can you break down this construction? Identify and examine them. How can you measure the density of poetic shocks thus created? Specify/appreciate the effect of poetic shocks.

Semantic indirection highlights two orientations: what one says and one actually means. What one says is that nature is a shrine but what factually means is that nature is a sacred place, inhabited by God. Hence arises the presence of two isotopies. An isotopy is, according the semiotician Umberto Eco, a same direction of interpretation of meaning within a text. Thus, there are two directions of interpretation of meaning: nature as an ecosystem (the first isotopy), nature as a divine sphere (second isotopy).

Poetic shocks: they are the product of the semantic distance between what one says and what one really means. Their density can be assessed by the extent of the semantic distance between referentiality (what one says: an <u>ecosystem</u>) and rhetoricity (what one factually means: <u>divine presence</u>). It appears that an ecosystem and God's presence are semantically far apart. Thus, juxtaposing these different notions nurtures a poetic shock of great magnitude and eminently honed by the semantic distance between nature and the temple. However, this poetic shock is 'assumed' by a semantic finding: the connection between nature and temple. God is in nature and the temple as well. He is their common denominator.

B—What is the literary implication of the expression "Nature is a temple"? What are its metaphysical implications ?

Since nature is inhabited by God, it bears a divine seal, that is perfection, pre-established harmony decreed by God. Its presence within nature disqualifies any form of randomness and leads to immanentism (that is a system of thought positing that God dwells in nature and operates within it). However, if God is in nature, He is also everywhere, which gives way to *pantheism (philosophical system highlighting God as being everywhere -> from Greek 'pan' = everywhere, all, and 'theos' = God). Pantheism is represented by the Latin formula "Deus sive natura" ("God or nature"), which was precisely the slogan of *Spinoza's pantheism: the view that god and nature are interchangeable, or that there is no distinction between the creator and the creation. Consequently, the expression "nature is a temple" nurtures two philosophical beliefs and systems: immanentism and pantheism. Pantheism also posits that every element of nature reflects divine perfection.

C—What do the constructions 'confused words', 'forêts de symboles' suggest?

'Confused paroles' suggests the notion of unity, harmony. With respect to "forêts de symboles" it suggests profusion, multitude and the idea of the mystery because the word 'symboles' by itself alludes to the mystery, unknown, and what is yet to be decrypted.

D—To what extent is the word 'unity' important in the second stanza? What does the 4th line of this stanza stand for? Which word nurtures the semantic matrix of the 4th line? How?

The word 'unity' is the semantic matrix that prepares the notion of synesthesia occurring in the 4th line of stanza # 2. The semantic matrix or nucleus is the focal point from which the meaning of a text/sentence and all its layers stem.

The word 'se repondent' nurtures the semantic matrix. It is a copula, the pre-requisite element of synesthesia. Here it is a verbal copula synthesizing three sensory modalities: 'parfums', 'couleurs', 'sons' and presiding over the creation of synesthesia through the psychological law of totality.

E—The title of the poem is 'Correspondences'. How does it help to deepen the understanding of the text. How many types of correspondences are there in the text?

'Correspondences' is a tacit reference to the connections between the material and spiritual as well as intra-cosmic unity. Therefore, it alludes to synesthesia, the main theme of the passage.

Identify them. In what stanzas do they occur?

They occur in Stanza # 3 and # 4.

F—The apex and aspiration of symbolism is Cratylism, that is, the attempt to reduce the arbitrariness of the signifier and, by the same token, motivate the connection between the signifier and the signified and, in the final analyis, maximize the power of language. How is Cratylism manifested in this poem?

APPLICATION OF SYNESTHESIAMETRICS TO THE POEM:

3 MAJOR CANONS/BENCHMARKS

1—Dynamic symbiosis between science and poetry/semiotics

In stanza # 2 the poem reaches its climax underscored by specific constructions. Identify them. The semantic matrix or semantic nucleus is the focal point from which the meaning of a text/sentence and all its layers stem. In light of this definition, show how the semantic matrix of these constructions can help reconcile science and poetry in this text? (unity, order -> 0 entropy -> 2nd law of thermodynamics: poetic explanation and vision of low/absence of entropy in the universe (beginning of the universe), science viewed artistically). Poetry and semiotics raise the question regarding the 'why' of unity/order in the universe whereas science provides the answer regarding 'the how' of this order. It establishes a causality principle. The why/cause -> poetry/art/semiotics, the because/effect -> science. Why is there logos (order, perfection, unity) in the cosmos (universe) ? Because of lack of chaos (disorder, entropy) => Entropy is 0. It is noteworthy that the word 'order' and science have the same etymological meaning. Here, they are interchangeable, and paradigmatically variants. Thus, they indistinctively means or at least refers to: '"science, study, order, knowledge, creative word, divine word, and speech" => *paradigmatic variants, are interchangeable lexical units vs. *syntagmatic units)

2—Semantic indirection assessed on the basis of the amount of poetic shocks

Idiolectic pertinence assessed on the basis of 3 criteria: poetic shocks, power of suggestion, and latent symbolism (the corollary of poetic shocks and power of suggestion)

Stanza # 2 is very important because the semantic matrix of the whole passage is embedded in it. Explore how the <u>semantic indirection is structured in this stanza to create poetic shocks</u> (semantic distance between 3 isotopies perfumes, colors, sounds

3—Combination of 3 semantic levels (literal & figurative, holistic, ontogenetic)

The metaphorical synesthesia can be found in the second stanza. (analysis of the structure: 3 inter-sensory isotopies/ modalities: perfumes (olfactory) + colors (visual) + sounds (auditory) + a verbal copula 'respond'.

<u>Note</u>: The power of poetic shocks is proportional to the semantic distance between "mimesis/referentiality" and "rhetoricity" as shown in this diagram. The word "mimesis" refers to the imitation of reality or referentiality (reality). By mimesis we mean what one says whereas "rhetoricity" stands for what one actually <u>means and the process employed to move from what one says to what one actually means</u>. The shift from mimesis to rhetoricity provides room for the exercise of semantic indirection.

II—"L'ALBATROS" by BAUDELAIRE

Souvent pour s'amuser les hommes d'équipage
Prennent des albatros, vastes oiseaux des mers
Qui suivent, indolents compagnons de voyage,
Le navire glissant sur des gouffres amers.

A Peine les ont-ils déposé sur les planches
Que ces rois de l'azur, maladroits et honteux
Laissent piteusement leurs grandes ailes blanches
Comme des avirons trainer à côté d'eux.
Ce voyageur ailé, comme il est gauche et veule!
Lui naguère si beau, qu'il est comique et laid!
L'un agace son bec avec un brûle-gueule,
L'autre mime en boîtant l'infirme qui volait!

Le Poète est semblable au prince des nuées
Qui hante la tempête et se rit de l'archer,
Exilé sur le sol au milieu des huées,
Ses ailes de géant l'empêchent de marcher.

EXPLORATION OF THE TEXT: 2 levels (See answers at the end of the book)

1—HERMENENEUTIC READING:

At the hermeneutic level, we read the passage for the first time and trt to understand its meaning. At this stage we come up with a number of interpretations. The following questions are intended to spark these interpretations:

QUESTIONS:

a—What are the nature, structure and main idea of this poem?

The passage is composed of 4 stanzas with homometrical verses. Each of them is an non-classical alexandrine verse in the sense that its rhythm is psychological rather than subject to conventional and metrical canons (see differences between classical, romantic poetry and symbolist poetry on pages 44-45).

Structure: two semantic units: the first covers the first three stanzas and can be titled: the capture of the albatross and his pathetic condition in the boat. The second covers the last stanza and can be titled: the identification of the albatross with the poet/genius

The albatross is a gorgeous bird when he is free in the sky, but when he is captured by sailors he becomes funny and subject to derision.

b—What does the periphrastic construction "hommes d'equipage" stand for? What is the meaning of "prennent". Define the word "albatros"

'Hommes d'equipage' stands for sailors, seamen, and, at a deeper level, common men, and the rabble, and those who lack culture, and critical analysis.

'Prennent' has the connotation of 'capturer' (to tame) it is a *euphemistic form of 'capturer'. Indeed, 'capturer' suggests violence, savagery with which aggressors capture an animal. The verb "prennent' is intended to linguistically 'soften' this aggressiveness.

The albatross is a charismatic bird. Its size is huge. He can have a wing span of more than three meters and a half in flight.

c—This poem epitomizes symbolism in terms of its theoretical program and esthetic canons. On the basis of your knowledge of symbolism, can you summarize the way it is presented in this text in 5 major canons? How? (suggestions for answers: idealism; music; phonic mimologism, synesthesia -> Cratylism; sophisticated vocabulary; free verses vs. freed verses => psychological rhythm vs. metrical rhythm)

The use of

- Synesthesia: gouffres amers = combination of a visual modality (gouffres) and an olfactory modality (amers)
- Symbols/allegory:

    • The albatross symbolizes the poet, the genius in the mist of the rabble and mediocrity;
    • The sailors represent the rabble, and mediocrity, the common man incapable of intellectual elevation.
    • The boat stands for the society, the unrefined men and their mediocrity.
    • His wings of giant ("ailes de geant") symbolizes the huge faculties of the genius

- Music: rich rhymes, psychological rhythm, asyndeton (stanza #1),
- Power of suggestion: suggestive lyricism and latent symbolism are rendered by a meticulous choice of words: periphrastic constructions/sophisticated vocabulary

(hommes d'équipage, rois de l'azur, prince des nuées,), vivid metaphors (ailes de géant, huées,), modifiers (Souvent, piteusement) metonymy (planches)
- Use of a psychological rhythm instead of a conventional/metrical rhythm

2—HEURISTIC READING

At the heuristic level, we do a second or even a third reading of the text, a retroactive one, which makes us to move to a deeper level and decrypt other layers of meanings. These questions will try to trigger deeper findings.

QUESTIONS

a—Identify synesthesia in the first line? Into how many isotopies can you break down this construction? Identify and examine them. How can you measure the density of poetic shocks thus created? Specify/appreciate the effect of poetic shocks.

"Gouffres amers" constitutes a synesthetic construction. Poetic shocks are nurtured by the semantic distance between 'gouffres' and 'amers' and enhanced by the semantic clash between them. A combination of these two words is not expected to occur in a verse or sentence because of their semantic incompatibility. Thus, their collision generates the esthetics of the surprise, adds a special feature to the whole construction. It follows that the lexical vicinity of the construction together with its paradigmatic and syntagmatic relationships exhumes its implicit meanings, which gives way to latent symbolism.

b—Symbolism:

What do these words stand for: hommes d'equipage, planches, rois de l'azur, prince des nuees? The word "infirme qui volait" introduces a contradiction. Explain this contradiction. Explain the construction "ses ailes de géant l'empechent de marcher".

"Hommes d'equipage" is a circumlocution, that is, a periphrastic construction and it refers to the crew, sailors, or just boatmen.

Planches" is a metonymy and designates the boat. The metonymy is a stylistic device in which a portion (planches is a part of the boat), an element is used for a whole set (the boat)

"rois de l'azur" and "prince des nuees" are periphrastic constructions designating the albatross. The words 'azur' and 'nuees' refer to clouds and figuratively to ethereal spheres where characters of high status are supposed to dwell. To a certain extent, these constructions are also metaphors highlighting the nobility of the albatross. He is, indeed, compared to a king ("rois de l'azur") and a prince ("prince des nuees").

The construction "l' infirme qui volait" outlines a paradox between the past and present. The past and present are coalesced in it. As a matter of fact, 'Infirme' highlights the current condition of the albatross: his crippled status in the boat. "Qui volait" highlights his past status: he was freely and graciously flying high in the air before being captured. On a different semantic level, the present stresses the current derision of the albatross, his pathetic aspect due to his inability to function in the new universe in which he is imprisoned. He is surrounded by people who represent the rabble and make fun of him. His past is glorious, stresses his nobility, superior faculties whereas the present is sad and even pathetic.

"ailes de geant" refers to the outstanding faculties, special assets that the genius possesses. Such faculties eventually turn out to be an impediment, a disadvantage because society is so intellectually limited that it cannot understand the magnitude of his genius and all its implications. Hence the use of the construction "l'empechent de marcher" that underlines the impossibility for the genius to fully express himself.

c—Is this poem allegorical or symbolical? Justify your answers in light of your knowledge of symbols and allegory. Explain its layers of meaning in light of the meaning of the albatross and that of "les planches"?

As you know, the aspiration of symbolism as poetic school is Cratylism. How is cratylism manifested in this poem?

In light of the definition of symbol and allegory, we can consider that this poem is both symbolical and allegorical. A symbol is a concrete representation of an abstract reality whereas an allegory is a development or accumulation of symbols often conceived for a didactic purpose. By virtue of this postulate, if we take each symbol of the poem individually we can consider that the text is symbolic. Additionally, there is one major symbol in the passage: the albatross and he stands for the poet, genius, magus, or even the wise man. However, if we take into account the fact that there are minor or peripheral/satellite symbols ('planches', 'hommes d'equipage', 'le sol' 'huees', 'ailes de geant' revolving around the major symbol, then, this poem becomes allegorical. Moreover, the allegory is usually didactic since it is conceived to convey a seminal lesson (e.g. Plato's *Allegory of the Cave*), a teaching, or express an aphoristic notion, that is a general truth that needs to be known, or at least reiterated/reminded. Precisely, this passage develops a central message, an aphoristic notion: the pathetic condition of the genius in society. He is ahead of his time. His faculties and talent are so high, sophisticatedly refined that his society becomes unable to understand or, at least, appreciate the magnitude of his genius. Therefore, society condemns or marginalizes him. Such was the case of Galileo, Copernicus, Baudelaire, Socrates sentenced to drink the hemlock because he was allegedly corrupting the Athenian youth, etc . . .

SYNESTHESIAMETRICS: 3 MAJOR CANONS/BENCHMARKS

- 1—*Possibility of dynamic symbiosis between neuroscience and poetry/semiotics
- 2—Semantic indirection assessed on the basis of the amount of poetic shocks
- 3—Idiolectic pertinence (power of suggestion + latent symbolism)
- 4—Two or more sensory modalities combining 3 semantic levels (literal, figurative, and ontogenetic)

This metrics can be re-written as follows:

Dynamic symbiosis between neuroscience and poetry/semiotics + semantic indirection assessed through the density of poetic shocks + combination of 3 semantic levels (literal, figurative, and ontogenetic)

Synesthesiametrics = Epistemological complexity x poetic quality x semantic density x idiolectic pertinence

*The first criterion (epistemological complexity) cannot be applied here because of lack of factual scientific data, which takes us to the other criteria:

- Poetic quality

    The poem has one synesthetic construction: 'gouffres amers'

    Poetic quality: Poetic shocks are nurtured by the semantic distance between 'gouffres' and 'amers' as well as the exercise of semantic indirection thus created, enhanced by the semantic clash between them. (see detailed answer in "a")

- Semantic density: coalition of a visual (gouffres) and gustatory modality (amers). The combination of their literal, figurative meanings added to their lexical vicinity ('Le navire glissant.") and paradigmatic and syntagmatic relationships create an ontogenetic meaning, that is a meaning generated

by itself, triggered by literal, figurative meaning and the aforementioned relationships
- Idiolectic pertinence: power of suggestion + latent symbolism

The power of suggestion and latent symbolism are expressed by a meticulous choice of words/constructions and the outstanding ability of the poet to collide them to create prolific meanings, those likely to generate themselves endlessly because they are multiplied by their semantic field (see chapter III). That is precisely the case of the choice of 'gouffres' and 'amers', the taking into account of their eminently suggestive effects, cornucopia of connotations, semantic ramifications conceived to generate symbols. When symbolic meaning prevails in a text, it also generates other symbols and, by the same token, other meanings, which leads to latent symbolism, and meaning re-created by an ontogenetic process.

III—LA VIE ANTERIEURE (Baudelaire)

J'ai souvent habité sous de vastes portiques
Que les soleils marins teignaient de mille feux,
Et que leurs grands piliers, droits et majestueux,
Rendaient pareils le soir, aux grottes basaltiques.

Les houles en roulant les images des cieux
Mêlaient d'une façon solennelle et mystique,
Les tout-puissants accords de leur riche musique
Aux couleurs du couchant réflété par mes yeux.

C'est là que j'ai vécu dans les voluptés calmes,
Au milieu de l'azur, des vagues, des splendeurs
Et des esclaves nus, tout imprégnés d'odeurs,

Qui me rafraîchissaient le front avec des palmes,
Et dont l'unique souci était d'appronfondir
Le secret douloureux qui me faisait languir

EXPLORATION OF THE TEXT: 2 levels

1—HERMENENEUTIC READING:

QUESTIONS:

a—What are the nature, structure and main idea of this poem?
b—This poem epitomizes symbolism in terms of its theoretical program and esthetic canons. On the basis of your knowledge of symbolism, can you summarize, in 5 major canons, the way it is presented in this text ? How?
c—The first two lines of the first quatrain refers to the setting. What information does the text provide regarding the characteristics of this setting? Examine lexical items and their semantic field to find this information. What is the stylistic device used in the second stanza? What does it suggest?

ANSWERS:

a—Nature: Sonnet composed of alexandrine verses, its rhythm is psychological, not canonical/physical.

Structure: 2 semantic units: a—the first two quatrains: emphatic description of a special universe where the poet has lived b—the two tercets: expression of the poet's vivid nostalgia

Main idea: Nostalgic expression of reminiscence -> this reminiscence reminds Plato's.

b—4 major canons of symbolism

    1—synesthesia: "les houles en roulant les images des cieux Melaient d'une facon solennelle et mystique, Les tout-puissants accords de leur riche musique Aux couleurs du couchant refletees par mes yeux"

Combination of tactile (roulant), visual (couleurs du couchant, images des cieux) and auditory sensory perceptions (accords de leur riche musique)

2—Use of music: special exploitation of recurrent phonic patterns: paranomasia (mystique/musique, cieux/yeux); imitative harmony with the sound [u] in h<u>ou</u>le, r<u>ou</u>lant, t<u>ou</u>t, c<u>ou</u>leurs, c<u>ou</u>chant); the sounds [oer] in l<u>eur</u>, coul<u>eur</u>

3—Use of sophisticated vocabulary + emphasis: tout-puissant accords de leur riche musique; vastes portiques, soleils marins, teignaient de mille feux (emphatic plural), piliers droits et majestueux, grottes basaltiques, solennelle et mystique, voluptés calmes (oxymoronic construction), au <u>milieu</u> de l'azur (strategic position and location (au milieu) in an ethereal universe (azur)), splendeurs (emphatic plural), languir, etc . . .

4—Cratylism: All the aforementioned considerations constitute an attempt to suggest a state of mind, to vividly express and re-create by the magic of language what the poet strongly feels by reducing the gap between the signifier and the signified and, in the same vein, trying to ward off the arbitrariness of language.

2—HEURISTIC READING

a—Symbiosis between science, emotion and poetry

- Physics: kinetic energy expressed by "dive into": a force deploys kinetic energy when it moves its center of gravity onto its straight line of action, which is translated by the equation: $EK = \frac{1}{2} m V^2$

- Emotion: manifested by an action verb emphasizing intensity = "to dive" enhanced by a pre-positioned adjective (<u>cold</u> darkness). In French an adjective can be pre-positioned or post-positioned. Here it

is pre-positioned (froides ténèbres) to underscore intense emotion, subjectivity, and lyricism
- Poetry/synesthetic metaphor: combination of two inter-sensory isotopies: froides -> isotopy of tactile sensory perception + isotopy of visual sensory perception.

b—The title "Vie antérieure" (Previous life) suggests a seminal concept and notion inherent in idealism and Platonism and symbolism. Explain it.
c—Which words or constructions nurture the semantic matrix leading to the generation of the whole poem?
d—The poem maximizes the use of music. What are the prosodic and literary implications of this process
e—What is the symbolism of the construction "esclaves nus"

1—Symbolism is based on Platonism and idealism. Platonism focuses on the ideal, supra-terrestrial world or world of archetypes. According to him, the soul was happy and used to live in this ideal word. Unfortunately because of a mistake committed, the soul was sentenced to live in prison (the body) in the terrestrial world. The Greek word 'soma'/'sema' designates both the prison and grave whereas 'psukke' refers to the soul, which explains the notion of a man as *psychosomatic* entity. Therefore, the soul is in exile in this material world but sometimes, he surreptitiously remembers scenes of his previous life. Some events triggered this vague remembrance which is precisely called 'reminiscence'. The whole poem revolves around this notion and superimposes the past (previous life in an ideal, ethereal world: 'azur') and the present (current life in the terrestrial world)
2—"Vie antérieure", "J'ai longtemps habité", "C'est là que j'ai vécu" -> the use of the passé composé (j'ai vécu) and the *deictic particle 'là' emphasizes the semantic nucleus
3—the outstanding use of prosody and recurrent phonic patterns lead to phonic mimologism and finally

to Cratylism because the poet aspires to create an equation between the gorgeous scene and the language used to suggest them. Therefore, the sophistication of language he uses is surmised to assess the sophistication of these gorgeous scenes.

4—'esclaves nus' ('naked slaves') symbolizes innocence, purity enhanced by the post-positioned adjective 'nus' because the post-position of adjectives has an objective connotation. It is noteworthy that here, this construction has a biblical connotation => in most sacred texts such as the Bible (especially The Old Testament), Baghavad Gita, Vedas, Upanishads, etc... nudity generally has a symbolic connotation: that of purity, innocence, candidness, or ingenuity.

APPLICATION OF SYNESTHESIAMETRICS

SYNESTHESIAMETRICS: 3 MAJOR CANONS/BENCHMARKS

1—Epistemological complexity: dynamic symbiosis between neuroscience and poetry/semiotics

The synesthesia: "Les houles en roulant les images des cieux Mêlaient les tout-puissant accords de leur riche misique Aux couleurs du couchant réflétées par mes yeux"

"Surges by rolling the images of skies Blended in a solemn and mystical way, The almighty chords of their rich music With the colors of the west reflected by my eyes"

Deploys a combination of an auditory perception (surges/ the almighty chords of their rich music) a tactile perception (rolling) and a visual perception (the images of skies). It triggers and inspires the idea of subtle and hidden connections likely to be found in the universe and bolstered up by science and arts (painting, poetry and semiotics). Indeed, since Einstein we have learned that energy and matter/mass are inter-convertible; $E = MC^2$, which shows the subtle connection between matter/mass and energy (matter being condensed energy). Just like mass and energy, light and sound are interrelated. Indeed, both light and sound/electromagnetic radiation can be viewed in one of two complementary ways: as a wave in an abstract electromagnetic field or as a stream of massless particles (photons). As waves, they are endowed with energy and their main differences can then be in terms of frequencies:

> Just as mass is condensed energy and waves expression of light and sound, here synesthesia is condensed information under the form of electrical signals channeled by synapses and neurons. They are subsequently transformed and interpreted by the thalamus as isotopies of colors/light/sight ('images of the skies', 'colors of the west reflected by my eyes'); those of sound/hearing (surges, almighty chords of their rich music); and touch (rolling). Finally, specialized psychic areas merge them and they become synesthesia by virtue of the psychological law of totality. That is precisely what we might realize in the modus of operandi of this synesthesia if we could analyze it scientifically.

2—Poetic quality The semantic indirection is assessed on the basis of the amount of poetic shocks

> The semantic distance between the auditory perceptions 'surge' and music' creates a poetic shock intensified by the verbal copula 'blended with' and the phrase 'almighty chords' suggesting amplified density; the other semantic distance between 'surges' (embedding an isotopy of noise) and 'rolling the images of skies' (embedding an isotopy of sight) is substantially enhanced by their semantic clash and the graphology of 'surges', 'images of skies', 'colors of the west' through their emphatic plurals. This additional semantic distance and clash generate one additional poetic shock whose power is proportional to the semantic distance between each isotopy. Such poetic shocks definitely endow the synesthesia with idiolectic pertinence and the ability to nurture Cratylism.

3—Semantic density

Combination of 3 semantic levels (literal & figurative, ontogenetic)

The synesthesia conjoins 3 levels of meanings: literal; figurative/holistic; and ontogenetic. The way synesthesia presents reality is literal but mixed with a sizable amount of sophistication due to tacit cross-modal comparisons, which makes it stand out. Thus, here, the perception has a literal meaning since the poet is supposed to express what he sees, hears and feels simutaneously: 'surges rolling the images of the skies'. To this literal meaning is added a figurative one, artistic and metaphorical because the art is conveyed by a metaphor: the coalition of music (surges are metaphorical vectors of music: 'the surges blended their almighty chords of their rich music); painting ('with the colors of the west reflected by my eyes', 'the 'sea suns' are metaphorical vectors for a canvas on which the painter is painting to distill the quintessence of his creativity inspired by the gorgeous spectacle of the sunset); and poetry (as an art is buttressed by the metaphorical synesthesia and the graphology of 'skies' since the plural use of 'skies' has a poetic and artistic connotation.). However, the literal meaning colludes with a figurative/artistic meaning. In fact, this synesthesia superimpose poetry, painting (images of the skies . . .) and music, and even choreography/ballet.

The ontogenetic level is induced by the lexical vicinity of inter-sensory modalities colluded to constitute the synesthesia: 'the surges by rolling the images of the skies, blended [. . .] the powerful chords of their rich music with the colors of the sunset'. 'The images of the skies', 'powerful chords', 'rich music', 'colors of the sunset' are lexical items combined to suggest the gorgeous beauty nurtured by the splendor of the sight (sunset), that of the music (powerful chords), the special and physical texture of the sea (surges . . . rolling). Their latent symbolism is designed in such a refined and sophisticated way that it suggests: painting, music, choreography, and ballet. Indeed, 'skies' emphatically used in the plural refers to the lexicon of painting and art. Music is stylistically articulated by 'powerful chords'. With respect to choreography and ballet, they are rendered by the morphological structure of the sea unfolding its elements (rolling surge, rolling waves, etc . . .) as if nature were involved in dancing. It follows that the symbiosis of painting, music, choreography, and ballet cogently supported by stylistic constructions used to express them endows this synesthesia with a powerful semantic density. All the lexical items used here are very suggestive. That is why they contribute to generate latent symbolism (interconnectedness of elements in nature, harmony, intra-cosmic unity, etc . . .) enhanced by the creation of underlying meanings which, with their semantic ramifications, confers upon the synesthesia the ability of producing meaning endlessly and motu proprio (that is by itself, on its own movement).

4—Idiolectic pertinence

It is rendered by the power of suggestion (music; suggestive lyricism: collusion and collision of especially expressive words or constructions) and latent symbolism (lexical vicinity added to semantic fields of words create layers of meaning (s) and tacit symbolism (houles, images des cieux, accords de leur riche musique)

IV-*CHANT D'AUTOMNE* by *BAUDELAIRE*

Bientot nous plongerons dans les froides ténèbres;
Adieu, vive clarté de nos étés trop courts!
J'entends déjà tomber avec des chocs funèbres,
Le bois retentissant sur le pavé des cours,

Tout l'hiver va rentrer dans mon être: colère,
Haine, frissons, horreur, labeur dur et forcé,
Et comme le soleil dans son enfer polaire,
Mon coeur ne sera plus qu'un bloc rouge et glacé.

J'écoute en frémissant chaque qui tombe;
L'échafaud qu'on bâtit n'a pas d'écho plus sourd
Mon esprit est pareil à la tour qui succombe
Sous les coups d'un bélier infatigable et lourd.

Il me semble, bercé par ce choc monotone,
Qu'on cloue en grande hâte un cercueil quelque part ...
Pour qui?. C'était hier l'été, voici l'automne!
Ce bruit mystérieux sonne comme un départ.

EXPLORATION OF THE TEXT:

1—HERMENEUTIC READING

    a—What is the nature of the poem? What is its main idea? Justify the title (elegy -> funeral song = from Greek 'elegeia' = 'funeral/grief song') -> explore the *lexico-semantic field of grief. Identify the stylistic device contained in the first line. Identify its components. What does it suggest?
    b—What is the structure of the poem?
    c—This poem epitomizes symbolism in terms of its theoretical program and esthetic canons. On the basis of your knowledge of symbolism, can you summarize, in 5 major canons, the way it is presented in this text ? How? (idealism; music; phonic mimologism, synesthesia -> Cratylism; sophisticated vocabulary; free verses vs. freed verses => psychological rhythm vs. metrical rhythm)

        a—Nature: poem of 4 quatrains with isometric verses and a rhyme scheme (ab, ab cd, cd, ef, ef, gh, gh) Main idea: anxiety of the poet when he senses the impending coming of winter.

        In the first line the poet uses a metaphorical synesthesia, the tone is pessimistic, dismal. It heralds a bad omen.

        Justification of elegy: the poem can be regarded as an elegy because it is infused with the symbolism of death, that of the summer and fall. Indeed the word 'elegy' means 'funeral song' in Greek ('elegeia'). Additionally, most words have an explicit or tacit semantic connection with funeral/loss/bereavement: *"adieu, chocs funebres, echafaud, tour qui succombe, cercueil, depart"*. The whole passage is built on a stylistic device called 'asyndeton', which expresses separation, disconnection. The asyndeton refers to the absence of logic connectors or suppression of any particle designating linking or connection.

Precisely, there are no such particles (or very little) in the text. Consequently, the poem functions as an elegy a departure, a separation of death from life.

b—Structure of the poem: 2 semantic units: quatrain 1 & 2: Premonition of pernicious events that will occur in the winter quatrain 3 & 4: the traumatic impact of these future events on the psyche and the physical of the poet. The poet feels them in anticipation.

c—canons of symbolism: idealism; music; phonic mimologism, synesthesia -> Cratylism; sophisticated vocabulary; free verses vs. freed verses

- Use of synesthesia: 'froides tenebres' => combination of a tactile (froides) and a visual sensory modality (tenebres) => idiolectic pertinence
- Use of music: sophisticated exploitation of recurrent patterns: homophones (cours/courts), minimal pairs (colere vs polaire [koler] vs. [poler]), (lourds vs. sourds [lur] vs. [sur], paronomasia (tenebres vs. funebres, part vs. depart), rich rhymes (. . . .) all these technical devices lead to phonic mimologism
- Psychological rhythm instead of a metrical rhythm. The psychological rhythm is exceptionally emphasized in stanzas 1 & 2
- Highly refined vocabulary and technical devices: collusion and collision of words meticulously chosen to produce a special effect (chocs funebres, bois retentissant avec des chocs funebres; colere/enfer polaire; mon esprit est pareil a la tour qui succombe (powerful simile obtained by the conjunction of "esprit" and "tour qui succombe" => creation of poetic shocks => idiolectic pertinence)
- Cratylism: Outstanding ability for the poet to reduce the gap between language and its factual reference, the signifier and the signified through a systematic tapping into all the aforementioned devices combined with lexical drunkenness (qualitative and quantitative use of very specific words leading to highly semantic density) and asyndeton (absence of logical connectors or most particles

likely to suggest linking). The poet focuses on the end of the fall and the impending coming of winter in a very vivid way. Thus, he talks about the "departure' of the fall, its 'death', that is the separation of these two seasons and their underlying psychological impacts. Through the language he utilizes, the reader can actually <u>sees</u> or even '<u>feels</u>' such separation with the asyndeton, phonic mimologism, and the whole linguistic arsenal employed by the poet. Consequently, there is little discrepancy or arbitrariness between the signifier and the signified. Linguistic signs definitely become motivated.

2—<u>HEURISTIC READING</u>

<u>Synesthesiametrics:</u>

a—Epistemological complexity: possible symbiosis between science, emotion and poetry

- Physics: kinetic energy expressed by "dive into": a force deploying kinetic energy when it moves its center of gravity onto its line of action . . .
- Emotion: manifested by an action verb emphasizing intensity = to dive enhanced by a pre-positioned adjective (<u>cold</u> darkness). In French an adjective can be pre-positioned or post-positioned. Here it is pre-positioned (<u>froides</u> tenebres) . . .
- Synesthetic metaphor: combination of two inter-sensory isotopies: froides -> isotopy of tactile sensory perception + isotopy of visual sensory perception

b—idiolectic pertinence

c—Σ of two or more inter-sensory modes combined with 3 semantic levels (literal, figurative, and ontogenetic)

"We will soon dive into cold darkness."

This metrics can be re-written as follows:

Dynamic symbiosis between neuroscience and poetry/semiotics + semantic indirection assessed on the basis of the amount or density of poetic shocks + combination of 3 semantic levels (literal, figurative, and ontogenetic)

Metrics of emotion = lyricism -> anything likely to suggest subjectivity/exaggeration, passion, lack of control: lyrical pronoun (me, I); prepositional adjectives; excessive use of adjectives; intensifiers/adverbs; proliferation of action verbs.

V—*SONNET DES VOYELLES* by ARTHUR RIMBAUD

A noir, E blanc, I rouge, U vert, O bleu: voyelles,
Je dirai quelque jour vos naissances latentes:
A, noir corset velu des mouches éclatantes
Qui bombinent autour des puanteurs cruelles,

Golfes d'ombre; E, candeur des vapeurs et des tentes,
Lances des glaciers fiers, rois blancs, frissons d'ombelles;
I, pourpres, sang craché, rire des lèvres belles
Dans la colère ou les ivresses pénitentes;

U, cycles, vibrements divins des mers virides,
Paix des pâtis semés d'animaux, paix des rides
Que l'alchimie imprime aux grands fronts studieux;

O, suprême Clairon plein des strideurs étranges,
Silence traversés des Mondes et des Anges:
—O l'Oméga, rayon violet de Ses Yeux !

A. Rimbaud

SYMBOLISM, SYNESTHESIA, AND SEMIOTICS, MULTIDISCIPLINARY APPROACH   269

I—HERMENEUTIC READING

a—what is the nature, structure and main idea of the text?
b—summarize three/four esthetic canons of symbolism

a—Sonnet built on a monolithic and semantic unity developed throughout the poem. The poem deals with the symbolism of vowels and colors on the basis of colored hearing.
b—1—synesthesia: it specifically focuses on color-grapheme synesthesia
   2—music: recurrent phonic patterns: imitative harmonies often created by the rhyme scheme (candeur vs. vapeur, tentes vs. penitentes, belles vs. ombelles, virides vs. rides, anges vs. étranges); alliterations in [v]: "vibrement . . . virides" in [p] "paix . . . patis . . . paix",

   In [s] and [tr] "suprême Clairon plein des strideurs étranges" "Silence traversés des Mondes"

   3—suggestive lyricism: special combination of words or constructions to suggest particular feelings, emotions, psychological landscapes,

II—HEURISTIC READING

Study the poetic quality, the semantic density, and
idiolectic pertinence of constructions in this sonnet
How does this poem attempt to achieve Cratylism?

Poetic quality: use of poetic shocks by semantic collision of words:

"A, noir corset velu" vs. "des mouches éclat antes";
"mouches éclatantes" vs. "qui bombinent";
"qui bombinent" vs. "puanteurs cruelles";
"Paix des patis" vs. "semes d'animaux"

Idiolectic pertinence -> meticulous choice of words or constructions
+ latent symbolism

—**Cratylism: The whole sonnet is constructed on the systematization of recurrent phonic patterns (alliterations, imitative harmonies are particularly maximized) leading to phonic mimologism, use of eminently suggestive words (inducing semantic collision), which gives way to lexical drunkenness:**

"A, **noir corset velu des mouches éclatantes**
Qui **bombinent** autour des **puanteurs cruelles**"

"**vibrements divins des mers virides,
Paix des pâtis semés d'animaux, paix des rides**"

"Lances des glaciers fiers, rois blancs, frissons d'ombelles;
I, pourpres, sang craché, rire des lèvres belles
Dans la colère ou les ivresses pénitentes"

All these considerations attest to the poet's aspiration to motivate language, to ward off its arbitrariness and, accordingly, establish an outstanding correlation between the signifier and the signified, which is precisely Cratylism.

VI—*THE SECOND COMING* by WILLIAM BUTLER YEATS

TURNING and turning in the widening gyre
The falcon cannot hear the falconer;
Things fall apart; the centre cannot hold;
Mere anarchy is loosed upon the world,
The blood-dimmed tide is loosed, and everywhere
The ceremony of innocence is drowned;
The best lack all conviction, while the worst
Are full of passionate intensity.

Surely some revelation is at hand;
Surely the Second Coming is at hand.
The Second Coming! Hardly are those words out
When a vast image out of *Spiritus Mundi*
Troubles my sight: somewhere in the sands of the desert
A shape with lion body and the head of a man,
A gaze blank and pitiless as the sun,
Is moving its slow thighs, while all about it
Reel shadows of the indignant desert birds.
The darkness drops again; but now I know
That twenty centuries of stony sleep
Were vexed to nightmare by a rocking cradle,
And what rough beast, its hour come round at last,
Slouches towards Bethlehem to be born?

HERMENEUTIC READING

1—a—Nature b—main idea of the poem

    a—Nature of the passage: Poem of free verses, no specific rhyme scheme, the rhythm is more psychological than physical or canonical
    b—Main idea of the text: Premonition/precognition of a new cosmic cycle of life
    c—Structure of the poem: 2 semantic units:

- Stanza # 1: cataclysmic vision/high level of entropy heralding the end of a cycle of life and the beginning of a new one;
- Stanza # 2: reiteration of the vision buttressed by a number of awe-inspiring precursor signs

    d—The title suggests the advent of a new cycle of life. From a religious perspective it suggests the arrival a second messiah, that of Christ, (which presupposes that he came once), or just a new savior. This perspective is supported by the fact that each cycle of life lasts 2000 years and requires a spiritual leader. Two passages from the poem endorse this idea: "twenty centuries of stony sleep", "its hour come round at last, /Slouches towards Bethlehem to be born". Indeed, "twenty centuries" is mathematically equated with two thousand years. With respect to "its hour come round at last, / Slouches towards Bethlehem to be born", it alludes to the location of Christ's birth.

2—Esthetic canons of symbolism

    <u>Symbols</u> -> The poem has several symbols: gyre, falcon and falconer, Second Coming, darkness, twenty centuries, rough beast

- Gyre: symbolizes the circle, the infinite (beginning -> end, beginning -> end, and son on). As a circle, it essentially stands for a palingenetic vision of the universe. Such a vision postulates that the universe complies with an endless repetition of cycles. This symbolism is emphasized by lexical items: the repetition of 'turning' (turning and turning), and "widening gyre". The repetition of "Surely . . . at hand", "Surely . . . at hand" reinforces this notion of cycle: beginning -> end and vice-versa. Here, the cycle is dominated by chaos, a high level of entropy.
- Falcon: in Christian semiotics, the falcon Christian symbol represents evil because the wild falcon is a predator that attacks other birds. It is also associated with conversion to the Christian faith from a pagan being to a civilized Christian being—casting aside his old way of life and takes on the form of a new man. The falcon also symbolizes freedom from slavery referring to a falcon which escapes from his master's hand and flies free into the wind. This last connotation seems to be suggested by the verse: "the falcon does not hear the falconer". If the falcon becomes free, it can no longer has any connection with the falconer. Therefore it returns to wild life, that is anarchy, chaos, entropy, disorder, which is expressed by the word 'anarchy'. Anarchy is also maximized by the fact that an animal leaving its master hones the notion of indiscipline, wild life, disorder rather than order. The presentation of stanzas 2, 3, 4, 5, 6 stylistically corroborate this perspective. Anyway, here, the falcon is much more associated with evil.

Stanzas 2, 3, 4, 5, 6 abide by a paratactic recurrent structure: each of these verses is built upon the patterns 'subject + verb' or 'subject + passive verb' and, to a significant extent, highlights a cycle dominated by chaos, anarchy.

<u>RECURRENCE OF A PARATACTIC STRUCTURE</u> ('V' stands for 'verse')

- V2 The falcon cannot hear the falconer (Subj + verb) => lack of order/chaos/anarchy

- V3 Things fall apart (Subj. + verb); the centre cannot hold (Subj. + Verb) => lack of order/chaos/anarchy
- V4 Mere anarchy is loosed upon the world (Subj. + passive verb) => lack of order/chaos/anarchy
- V5 The blood-dimmed tide is loosed (Subj. + passive verb) => lack of order/chaos/anarchy
- V6 The ceremony of innocence is drowned (Subj. + passive verb) => lack of order/chaos/anarchy

- "Second Coming" stands for repetition, a cyclical process
- 'Darkness' symbolizes ignorance, evil and its forces because ignorance is often nurtured by ignorance. Socrates pertinently used to mention: "nobody is wicked willingly" because wickedness, that is, evil, is caused by ignorance. By extension, darkness refers to the forces of destruction and chaos as well.
- "Twenty centuries" represent a new cosmic cycle because, according to most traditions, each cycle is supposed to last two thousand years (100 years time 20 makes 2000 years).
- The "rough beast" stands for the Anti-Christ. The poem heralds the coming of Christ or a new messiah. Precisely, most biblical texts consider that this arrival is preceded by that of the Anti-Christ. So, the Anti-Christ is emblematic of Pre-Apocalyptic realities. It undergirds the impending Christ's Second Coming from a Christian viewpoint and another messiah or avatar from a mystical viewpoint. Either of them is supposed to come and redeem mankind or, at least, lead it to a golden age of spirituality.

All these symbols nurture an allegory since an allegory is, by definition, a development/accumulation of symbols in a text usually presented poetically or philosophically to express a didactic, aphoristic, cosmogonical (e.g. *The Allegory of the Cave* by Plato), or even cosmological function. Precisely, here, the allegory is presented poetically with an implicit didactic function: the recourse to spiritual and moral values to save the world.

Idiolectic pertinence = power of suggestion + latent symbolism

The power of suggestion and latent symbolism are achieved through the judicious choice of words or constructions endowed with a highly suggestive effect and multiplied by their semantic field and lexical vicinity, which carries this effect a step farther:

"The ceremony of innocence is *drowned*" -> the word 'drowned' radicalizes the extent of the moral crisis

"Things *fall apart*; the *centre cannot hold*; *Mere anarchy* is loosed upon the world"; the blood-dimmed tide is loosed."

"Things fall apart", "Center cannot hold", "Mere anarchy" constitute a very eloquent illustration of this deleterious atmosphere of crisis. Society no longer has solid moral foundations ("the center cannot hold"). There is systematized chaos stressed by the use the modifier 'Mere' ("Mere anarchy") and a tremendous sense of havoc and loss of rationality ("Things *fall apart*") that is taken a step farther with the adjunction of "the blood-dimmed tide is loosed". Society has lost its centripetal forces, that is: moral guidelines and now generates centrifugal forces, those that nurture chaos, havoc, destruction, and moral deliquescence. It factually reaches the paroxysmal point translated into these terms: "The blood-dimmed tide is loosed and *everywhere* /The ceremony of innocence is *drowned* ". Two (2) key-words eminently materialize this acme of the crisis: *'everywhere'* and *'drowned'*.

Periphrastic constructions and tropes:

"rough beast" refers to the Anti-Christ

"A shape with lion body and the head of a man" suggests the sphinx, a mythical creature half man, half lion stemming from Egypt, Greek and Syria. Here, it seems to represent the Egyptian one because of its lexical vicinity with 'desert'. Unlike the Assyrian and the Greek ones, it does not have wings, which, apparently, does not symbolize any divine origin or kinship because wings symbolize immortality, divine creatures (angels, higher entities), and, to a certain extent, aspiration to transcendence. The sphinx, as a striking epitome of conundrum and opaque mystery, has been known for thousands of years for its bearing secrets and terrifying aspect. It thus, stands for a graphic combination of threat and mystery.

"a rocking cradle" is a metonymy for Jesus. Here, the container (the cradle) stands for the content (Jesus as a baby)

—<u>Music</u>: recurrence of phonic patterns are conceived to create musicality

- Repetition:

    "Turning and turning in the widening gyre"
    "The Second Coming"

- Alliterations in [s]:

    "Troubles my sight: somewhere in sands of the desert"
    "The ceremony of innocence"

- Anaphoric structures:

    "Surely some revelation is at hand;
    Surely the Second Coming is at hand"

—<u>Free verses</u>:

Symbolism maximizes the use of free verses to avoid inserting creativity into an orthodox straightjacket. Thus, free verses are intended to free imagination and catalyze innovation. That is precisely what the poet chooses to use throughout his text.

B—<u>HEURISTIC READING</u>

Semantic nucleus/matrix The construction "widening gyre" is the semantic nucleus because it suggests the concept of cycle or repetition, which is precisely the notion out of which the whole text builds its meaning and underlying meanings.

The first stanza is a powerful expression of chaos. How is this chaos suggested stylistically? Chaos is suggested by:

- A very vivid expression of crises: moral crisis, crisis of confidence and trust ("the best lack all conviction"), egregious behavior and deleterious action ("the worst are full of passionate intensity"), which leads to chaos.
- Structural complexity of constructions in Stanza 2: Antithesis + hyperbole:

    • Antithesis: "The best lack all conviction" is opposed to "the worst Are full of passionate intensity".
    • Hyperbole: The worst are full of passionate intensity"

What is the meaning of "spiritus mundi"? How is this meaning related to the symbolist conception of the poet?

There is a seminal isotopy in the passage. Identify it. What words or constructions endorse this isotopy?

- The "spiritus mundi" stems from Latin and means "the spirit of the world ". This construction refers to the Akashic Records also known as the universal memory which is part of the Cosmic Consciousness, that is, the ethereal Consciousness stemming from the divine and bearing all the data of human and Cosmic soul. Yeats did believe that each human mind is linked to a single vast intelligence, and that this intelligence causes certain universal symbols to appear in individual minds.
- The deep meaning of "Spiritus Mundi" shows that every human being is connected to the Universal Intelligence that is also called God, the Great Architect or Cosmic. It is possible to merge with this Universal Intelligence through spirituality: intuition, inspiration, illumination, meditation, and prayer. That is precisely one of the goals of symbolist poetry. Indeed, the real symbolist poet is viewed as a seer, visionary, or clairvoyant endowed with the ability to attune with the divine through poetry and serve as a privileged vessel between the ethereal and the material. Such a privileged status enables him/her to decrypt messages from the supra-terrestrial and translate them into poetry

through synesthesia, correspondences, metaphors, symbols, and allegory. Similarly, establishing a fruitful attunement with the Cosmic Consciousness enables the poet to see the future, or have intuition and inspiration, which grants him access to higher truths. Poetic inspiration, intuition are the mystical and technical devices that hone the asset of the symbolist poet, which is factually the asset of the poet demonstrated in this passage through vision and precognition (particularly in stanza # 2: "a vast image out of Spiritus Mundi", the "rough beast").

- The seminal isotopy subsumed in the poem is that of the circle or cyclical return. Words and constructions endorsing it are: "Turning and turning", "widening gyre", "Second Coming "(twice). The word 'gyre' from Latin 'gyrus': 'circle' with its modifier 'widening' particularly underscores this isotopy.
- Contribution of this passage to modern literature

Through this text, the poet highlights the use of free verses and, accordingly, the technical possibility and advantage to enhance freedom and creativity. It is the legacy that a sizable number of symbolist poets have passed down onto modern literature.

- Cratylism:

Is manifested by means of a sophisticated language by means of which the poet endeavors to reduce the gap between the signifier and signified: meticulous choice of words, psychological rhyme, free verses maximizing poetic creativity, vivid expressions, musicality (recurrent pattern: alliterations, anaphoric structures, etc . . .)

VII—*CIMETIERE MARIN* by PAUL VALERY

Ce ciel tranquille, où marchent des colombes,
Entre les pins palpite, entre les tombes;
Midi le juste y composée de feux
La mer, la mer, toujours recommencée!
O recompense apres une pensée
Qu'un long regard sur le calme des dieux!

HERMENEUTIC READING

- Nature, main idea of the poem, and structure:
- Nature: Excerpt of 6 *decasyllabic verses whose rhyme scheme is (aa) (bc) (cb)—Main idea: Remarkable peace catalyzed by *telluric elements and infused with vitality.
- Structure: coherent semantic unit expressing the notion of peace and vitality
- **Since this poem expresses the notion of peace and vitality, what lexical items suggest these notions?**

A—Peace

- The post-positioned adjective 'tranquille': the post-position of adjectives has an objective connotation. So, here, it emphasizes the notion of peace, quietness
- 'Colombes', which means 'dove' symbolizes peace, harmony, and even beauty and love
- 'Le calme des dieux': 'dieux' or gods suggests perfection or cosmic entities who have reached perfection -> 'le calme des dieux' relates to a kind of quietness inherent in those who have achieved perfection, absolute calm. It also refers to the very quietness of nature (the sky, sea, etc . . .)

B—Vitality

- 'palpite' ('Ce ciel . . . palpite') personification/metaphor in absentia highlighting stamina, energy suggested by 'palpite' (throbbing);
- 'tombes' (grave, tomb) is a place of peace, where one is supposed to have eternal rest especially after a laborious and productive life. Hence stems the last part of the epitaph that is generally written on tombs: "requiescat in pace" meaning "rest in peace".
- 'Feux' (fire) is a powerful expression of liveliness and vitality. It is with water, air, and earth, one of the four alchemical elements. As such, it is likely to transform and purify objects;
- 'La mer, la mer' (the sea) is one of the most eloquent expression of life and vitality because the sea is composed of water and water stands for life itself. Indeed, la mer nurtures life constantly,

endlessly. That is why it symbolizes the infinite. Such symbolism is enhanced by the repetition 'La mer, la mer'

C—3 or 4 symbolist esthetic canons

<u>Symbols</u>:

'Colombes' (doves): symbolizes peace, harmony, beauty, and, to a certain extent, love.

'Tombes' (grave): symbolizes peace and eternal rest

'Midi le juste' (Noon the fair, the just): symbolizes justice, equality. Indeed, at noon, the sun reaches the zenith and divides the day into two equal parts: 12 hours and 12 more hours. Therefore, midi (noon) stands for the Perfect Being, epitomizing a great sense of justice.

'Feux' (fire): represents the sun and vitality

<u>Music</u>:

Recurrent phonic pattern used to produce special effects:

- Rich rhymes (col*ombes* vs. t*ombes*, f*eux* vs. adi*eux*, recomm*encee* vs. p*ensee*)
- *La mer, la mer*
- *Entre* les pins palpite, *entre* les tombes
- Alliterations in [p] "les pins palpite", "O recompense apres une pensee" in [t] "Entre . . . palpite, entre les tombes"

<u>Idiolectic pertinence</u> = power of suggestion + latent symbolism

The power of suggestion and latent symbolism stem from the use of symbols, music, and the minute choice of words/constructions and the outstanding ability of the poet to collide them to create prolific meanings, that is, meanings likely to generate themselves endlessly because they are multiplied by their semantic field (see chapter III). That is precisely the case of the choice of 'le ciel . . .' colliding with '. . . palpite', the taking into account of their eminently suggestive effects, cornucopia of connotations, semantic ramifications conceived to generate symbols.

HEURISTIC READING

Semantic nucleus is nurtured by the isotopy of quietness that can be decrypted in these specific words and constructions: "ciel tranquille, colombes, le calme des dieux". All of them jointly contribute to create the concept of calm tacitly rendered by the title itself "Cimetiere marin". The whole text is built out of this seminal isotopy.

*Intertextuality

It is possible to find connections between this passage and Mallarmé's *Brise Marine* in terms of main theme, isotopy, aspiration to the ideal world (a world in which there is adamantine peace and bliss. *Cimetiere Marin* outlines such peace => hence the thematic connection between Mallarmé's *Brise Marine* and Valery's *Cimetiere Marin)*

VIII—*BRISE MARINE* by Stephane Mallarmé

La chair est triste, hélas! et j'ai lu tous les livres.
Fuir! là-bas fuir! Je sens que des oiseaux sont ivres
D'être parmi l'écume inconnue et les cieux!
Rien, ni les vieux jardins reflétés par les yeux
Ne retiendra ce coeur qui dans la mer se trempe.
O nuits! ni la clarté déserte de ma lampe

Sur le vide papier que la blancheur défend
Et ni la jeune femme allaitant son enfant.
Je partirai! Steamer balancant ta mature,
Levé l'ancre pour une exotique nature!
Un Ennui, désolé par les cruels espoirs,
Croit encore à l'adieu suprême des mouchoirs!
Et, peut-être, les mâts, invitant les orages,
Sont-ils de ceux qu'un vent penche sur les naufrages.
Perdus, sans mâts, sans mâts, ni fertile îlots.
Mais, O mon coeur, entends le chant des mâtelots!

QUESTIONS:

HERMENEUTIC READING

- Indicate the nature, main idea and structure of the passage.
- Poem of 16 non classical alexandrine verses with flat rhymes (aa, bb, cc, dd, ee, ff, gg, hh), focusing on the inner aspiration to leave for an ideal world represented in the text by "l'ecume inconnue "and "les cieux".
- Monolithic and semantic unit deploying the visceral tension of the poet towards an ethereal world.
- Summary of the text in 3 major esthetic canons of symbolism

1—Music

Rendered by a very judicious use of suprasegmental features, musical elements: paranomasia molded by rich rhymes (livres/ivres -> v 1, v 2; cieux/yeux -> v 3, v 4; defend/enfant -> v 7, v 8), minimal pairs (mature/nature -> v 9, v 10); repetition (Fuir! ... fuir -> v2) (perdus, sans mats, sans mats -> v 15); incantation molded by vocatives (O nuits! -> v 6) O mon coeur! -> v 16)

2—Usage of symbols

Blancheur, les cieux, l'écume inconnue, l'Ennui

"les cieux" symbolizes the ideal world, blancheur stands for purity, candidness, sincerity; "l'écume inconnue" represents poetic adventure; "l'Ennui" is a personified abstraction/allegory standing for the material world with its paraphernalia of frustrations and tribulations as opposed to the ethereal world.

3—Suggestive lyricism is rendered by a psychological rhythm instead of a canonical or physical one.

HEURISTIC READING

Intertextuality: the poem has a connection with a number of Baudelaire's poems and especially *Invitation au Voyage, La Mort des Pauvres, Le Vin des Amants*. The unifying theme of these three poems is the aspiration for the poet to leave this world for an ideal one. Mallarmé, who was Baudelaire's admirer and, to a certain extent, a disciple, has certainly drawn his inspiration from him. The tone of all these poems is similar because it is marked by the disillusionment and yearning to materialize the unique aspiration of the poet.

Cratylism is expressed by:

- A highly refined lexical collusion to create or suggest very special effects: "la chair est triste", le coeur . . . se trempe", "la clarte deserte de ma lampe", "l'adieu supreme des mouchoirs", "les mats invitant les orages", oxymoronic allusion ("cruels espoirs"),;
- Phonic mimologism (see music)
- Asyndeton (a stylistic device featuring the absence of connectors and connection) can be found in this passage. De facto, most verses are characterized by the absence of connectors such as "and, or, but, for, because, etc" . . . However, there are a few cases where connectors are used, but most of them are pseudo-connectors. For instance 'et' in "<u>et</u> j'ai lu tous les livres", "<u>Et</u>, peut-etre, les mats, invitant les orages" connect nothing. They do not function as connectors but as particles maximizing emphasis rather than any kind of connection. Consequently, the use of the asyndeton is conceived to suggest the separation aspiration to depart, to leave the terrestrial world for the ethereal world and reaching its peak in the repetition: "fuir! . . . la-bas fuir!) (meaning "Run! Run away!). It follows that asyndeton, phonic mimologism, and the highly refined lexical collusion are precisely part of the linguistic arsenal and the semiotic process judiciously exploited by the poet. They are designed to ward off the arbitrariness of signifiers/to motivate poetic language and, by the same token, epitomize Cratylism.

IX—SHORT EXCERPT FROM *Bruges-La-Morte*

Une flamme lui chanta aux Oreilles.
Un picotement lui vint aux yeux.
Il sentit un brouillard contagieux lui entrer dans l'âme;
envahi par le silence froid.

I—HERMENEUTIC READING

- Nature:

    Excerpt from *Bruges-La-Morte*, a poetic prose novel by Rodenbach, highlighting the Decadent esthetics through a very special use of language and synesthesia coined "style artiste"

- Structure and main idea:

    Semantic unit in which the narrator describes the eerie psycho-physiological landscape of the protagonist, Viane who mourns for his beloved. The weight of the loss has induced a highly complex state of passiveness, which causes him to behave in a very strange way. Precisely, this short passage is conceived to show the complexity of his behavior. As we mentioned in chapter 3, this text develops 'synesthesia en style artiste'. It has an eminently sophisticated status, a high level of structural refinement including two or three inter-sensory isotopies in which a passive verb or that which has a passive connotation is inserted; a personal pronoun ('lui') placed between the verb and the noun. Such is the idiolectic pertinence that features 'synesthesiae en style artiste'. Here is a few pertinent instances:

    Une flamme lui <u>chanta</u> aux Oreilles. Un picotement lui <u>vint</u> aux yeux

II—HEURISTIC READING

Systematization of paratactic structures

Each synesthetic sign is structurally identical to others because it is made up of: two or three inter-sensory isotopies in which a passive verb or that which has a passive connotation is inserted; a personal pronoun ('lui') placed between the verb and the noun. The verb usually functions like a deponent verb, that is, a verb with a passive signifier and an active signified (however, here the verb has a passive connotation). Such verbs are usually found in Latin. De facto, All these constructions are paratactic. Here is a few pertinent instances:

Une flamme	*lui*	*chanta aux Oreilles*	*(1). Un picotement*	*lui*	*vint*	*aux yeux (2)*.
Noun	Pron.	P. Verb	Noun	Pron.	P. Verb	

'Pron.' stands for 'pronoun', 'P. Verb' stands for 'passive verb'

Break-down of synesthesia # 1

(1) A visual inter-sensory isotopy (flamme) + an auditory inter-sensory isotopy (chanta aux oreilles)

Break-down of synesthesia # 2

(2) A tactile inter-sensory isotopy (picotement) + a visual inter-sensory isotopy (vint aux yeux)

<u>**Semiotic implications of paratactic structures**</u>

These paratactic structures of synesthesia "en style artiste" are conceived to suggest the particular psychological landscape of the protagonist. On the basis of their morphology (passive verb and pronoun placed between a passive verb and a noun), we can say that they suggest the isotopy of passiveness, numbness or insensitivity of the protagonist to anything surrounding him. As such, they reveal his disconnection from the outside world and his proneness to be <u>permeated</u> by external factors. As a matter of fact, Viane has lost his beloved. He is so enshrouded by his grief that he becomes totally immersed into inaction, passiveness. De facto, the weight of grief is so heavy that it has numbed his soul by eroding emotions, feelings and resilience from him. He has become like an aged man so weak that he is predisposed to collapse. Hence this quaint psychological state that is semiotically underscored by the special structure of the synesthesia (two inter-sensory isotopies between which a pronoun is inserted [lui]) and its corresponding isotopy: passiveness, numbness, weakness and, more exactly, quasi-decline. In this prose poetic novel, such a state is often expressed by the adjunction of an affix (inegayees, inapitoyees) conveying the idea of suppression, ablation, declension, or fall; sentences, lines, verses or constructions in which the narrator transforms an abstract force into a concrete force to make it act on beings ("Il sentit un brouillard lui entrer dans l'ame"). All these stylistic features are emblematic of the Decadent esthetics inherent in the Decadent movement, and resulting from a schism that was created within the symbolist movement circa the end of the nineteenth-century. One of the *iterative motifs of Decadent texts is ageing or senescence. Ageing or senescence strikes beings and acts on them in such a way that they are immersed into a state of passiveness marked with complacency. Such is the idiolectic pertinence that hones 'synesthesia en style artiste'. It can be examined at a deeper level through the isotopy of numbness and fall.

Isotopy of numbness and fall at a deeper level: premonition of modernity

In *Bruges-La-Morte* synesthesia, as an expression reflecting the modus operandi of the universe, is characterized by an internal paradox. Indeed, synesthesia is designed to compensate or ward off the lack of order of the universe caused by entropy by restoring order, unity. Thus, order must be 'recreated' from disorder, which gives birth to centripetal forces or forces of order as opposed to centrifugal forces or forces of disorder. Within the framework of all these considerations, synesthesia en style artiste stands for centrifugal forces. They need to be governed or compensated by other forms of synesthesia (catachrestic, metonymic, or metaphorical synesthesia), those nurturing centripetal forces, and not subscribing to disorder. Consequently, in *Bruges-La-Morte*, the use of synesthesia en style artiste is conceived to suggest the decline of the universe and its loss of focus. Indeed, in Viane's universe, the meaning exponentially escapes, everything seems to collapse. This trend to chaos and meaninglessness heralds modern literature. Precisely, Viane's universe reflects that of modern literature at a microcosmic level. Factually, this corpus where there is a constant process of semantic evasion bears the seal of modern literature and even harbingers Derrida's post-structuralism. Viane himself is flummoxed because he aspires to a coherent understanding of the meaning of life and whatever happens to him. His ego is sick, prone to collapse, which is suggested by the use of synesthesia en style artiste and its underlying isotopy. However, the emergence of other types of synesthesia tacitly suggests the necessity to attempt to re-unify the universe and his own by endowing it with a coherent meaning, a purpose and a form of order. Such a modus faciendi is regulated by metonymic, metaphorical, or catachrestic synesthesia ("a voice of the same color", a "colorless silence". Additionally, The decline of the universe is expressed by constructions composed of affixes suggesting the fall, collapse or ablation: 'in—' or 'dé—'. In *Rodenbach Les Décors du Silence* Patrick Laud provides an interesting analysis of this metaphysical phenomenon and presents a compelling catalog of affixes suggesting the decline, crumbling or fragility of the universe. He states:

> Dans le même régistre d'imagination décadente, on ne peut manquer de faire référence au substantif "émiettement", si caractéristique du monde de Rodenbach. Le verbe "s'émietter" revient lui aussi régulièrement pour exprimer de façon on particulièrement expressive l'éclatement et la dispersion symboliste: la fragilité de la "croûte" du monde, c'est l'absence d'unité qui caractérise au fond la structure du monde. Et on peut parler chez le décadent d'un véritable vertige de la multiplicité fragmentée. Cette fragmentation, cette absence d'une architectonique du sens constitue sans aucun doute l'aurore de la modernité. Il est d'ailleurs une caractéristique du lexique de Rodenbach qui prend en compte, mieux que toute autre de cet émiettement du monde décadent. Il s'agit de l'usage massif de verbes et d'adjectifs formés sur le préfixe 'dé—' et par là même expressif d'une perte de la forme [. . .]: dédoré, démodé, desséché, défraîchi dénudé, désenfiler se déflorer, se déprendre, se déclorer, se désagréger, déajuster, détroner, dévelouté, déteint, décalqué, se décomposer, se dénouer, se desserrer, déserter, décolorer, se dénuancer . . . Comment mieux exprimer cette "décadence" tous azimuts qui semble marquer la réalité? Cette décadence n'est, en fait, rien d'autre que l' éclatement ou l'effacement—suivant

le cas—de la surface formelle de l'existence, son "émiettement", qui n'est pas seulement multiplication et désarticulation, mais aussi épiphanie de l'informel et du "néant" (98).

Consequently, it calls for a philosophical reflection on the very essence of existence. Such is the deeper reading of this short text and its synesthesia en style artiste. Its letter and spirit, (through the expression of a complex/paradoxical sense of dislocation and unity) herald modern literature but also manifest themselves as an epistemological curiosity.

SELECTED BIBLIOGRAPHY

Books on symbols and symbolism

Johansen, Svend. *Le Symbolisme*. Paris: PUF, 1990. Print.
Nozedar, Adele. *Signs and Symbols Sourcebook*. New York: Metro Books, 2008. Print.
O'Connel, Mark et al. *The Ilustrated Encyclopedia of Symbols and Signs*. New York: Metro Books, 2007. Print.
Peyre, Henri. *Qu'est-ce que le Symbolisme?* Paris: PUF, 1974. Print.

Books and articles on synesthesia:

Cytowic, Richard. *Synesthesia: A Union of the Senses*. New York: Springer-Verlag, Inc, 1989. Print.
Duffy, Patricia. *How Synesthetes Color their World*. Oxford: New York, 2000, Print.
Harrison, John and Baron-Cohen, Simon. *Synaesthesia: Classical and Contemporary Readings*. Cambridge, MA: Blackwell Publishers Ltd, 1997. Print.
Heyrman, Hugo. "Art and Synesthesia: in search of the synesthetic experience", Lecture presented at the First International Conference on Art and Synesthesia in 2005, Print.
Ramachandran, Vs et al. *Synesthesia, Perspectives from Cognitive Neuroscience*. Oxford University Press, New York, 2005. Print.
Van Campen, Cretien. *The Hidden Sense: Synesthesia in Art and Science*. Cambridge, Massachusetts, The MIT Press, 2010. Print.

Books on semiotics, semantics, stylistics, linguistics and literature:

Barthes, Roland et al. *Sémantique de la Poésie*. Paris: Points, 1992. Print.
—*L'Obvie et l'Obtus*. Paris: Seuil, 1975. Print.
Genette, Gerard. *Figures II*. Paris: PUF, 1988. Print,
Kristeva, Julia. *Essais de Sémiotique*. La Hague: Mouton, 1971. Print.
Patillon, Michel. *Précis d'Analyse Littéraire*. Paris: Nathan, 1977. Print.
    Rey-Debove, Josette. *Lexique Sémiotique*. Paris: P.U.F, 1979. Print.
Riffaterre, Michael. *Semiotics of Poetry. Indiana:* Indiana University Press, 1978. Print.

Books and articles on emotion and intelligence

Thorndike, E. L. *Intelligence and Its Uses*, (Harper's Magazine 140, 227-335). Print.

Articles on symbolists

Williams, Adelia. *Verbal Meets Visual: An Overview of poésie critique at the fin-de-siècle. French Review.* Vol. 73, No 3, February 2000. Print.

Books on Rodenbach, Mallarmé, Symbolism and Surrealism

Baudelaire Charles. *Les Fleurs du Mal*. Paris: Hachette, 1996. Print.
Breton, André. *Second Manifeste du Surréalisme*. Paris: Jean-Jacques Pauvert, 1962. Print.
Gray, Christopher. *Cubist Aesthetic Theories,* Baltimore: Johns Hopkins Press, 1953. Print.
Greimas, Algirdas Julien. *La sémantique structurale: Recherche d'une méthode*. Paris: Larousse, 1987. Print.
Laude, Patrick. *Rodenbach, Les Décors du Silence*. Brussels: Editions Labor, 1990. Print.

Books on music therapy, articles on therapy, and deep breathing

Chopra, Deepak. *Beyond the Five senses: the Healing Power of Meditation* in Going Bonkers? Magazine. Kaly, Texas, October issue, 2009. Print.

Davis, Gfeller, Thaut, (2008). *An Introduction to Music Therapy Theory and Practice*-Third Edition: The Music Therapy Treatment Process. Silver Spring, Maryland. pg. 475. Print.

Kitab al-Musiqa al-Kabir. Trans. Ben Judah, Joseph ibn Aknin Al-Farabi's Kitab al-Musiqa al—Kabir. Print.

Roth, Edward. *"Neurologic Music Therapy"*. Academy of Neurologica Music Therapists Western Michigan University. 19 April 2011. Print.

Vey, Gary. *The Pineal Gland: the Seat of the Soul (Hope and the Pineal Gland through Breathing and Chanting)* Medford, OR: Winter Ventures, 2010. Print.

Books on Neuroscience, Cosmology, and Science in general:

Aertsen, A. *Brain Theory.* Elsevier Science publishers B.V., 1993. Print.

Gell-Mann, Murray. *The Quark and the Jaguar.* New York: Henry Holt and Company LLC. Print. 1994.

Gershon, Michael. *The Second Brain.* New York: HarperCollins, 1999. Print.

Greene, Brian. *The Elegant Universe.* New York: First Vintage Books, 2000. Print.

Kaku, Michio. *Hyperspace.* New York: Anchor Books, 1995. Print.

Robinson, Andrew. Einstein, A Hundred Years of Relativity. New York: Metro Books, 2010. Print.

Scientists and Contributors. *The Encyclopedic Dictionary of Sciences,* Oxford, New York, 1999. Print.

Reference books: Encyclopediae and Specialized Dictionaries

Cuddon, J.A. Dictionary of Literary Terms and Literary Theory. New York: Penguin Books, 1995. Print.

Rohmann, Chris. *Dictionary of Important Theories, Concepts, Beliefs, and Thinkers.* New York: Ballantine Books, 1999. Print

Preminger, Alex et al. *The Princeton Handbook of Poetry Terms.* New Jersey: Princeton University Press, 1986. Print.

Books on Literary History

Lagarde, André et Michard Laurent. *Dix-Neuvième Siècle*. Paris: Bordas, 1964. Print.
—*Vingtième Siècle*. Paris: Bordas, 1964. Print.

Books on Philosophy, Life, Math, and Broad Knowledge

Aristote. *Ethique à Nicomaque*. Trans. Jean Tricot. Paris: Vrin, 2002. Print.
Platon. *La République*. Trans. Victor Cousin. Vol. 7. Paris: Rey et Gravier, 1834. Print.
Rampa, Lobsang. *Chapters of Life*. London: Corgi Books, 1967. Print.
—*You Forever*. Boston, MA: Weiser Books, 1990. Print.
Solomon, Robert. *The Little Book of Mathematical Principles, Theories, and Things*. New York: Metro Books, 2008. Print.

INDEX

INDEX OF NOTIONS, CONCEPTS, & IDEAS   Pages

Adynaton synesthesia	136
Algorithm in semiotics and synesthesia	137
Anacoluthon	167
Apophantic function	130
Asyndeton (singular), asyndeta (plural)	59, 67, 285
Booba/Kiki experiment	172
Brainwave entrainment	188
Catachresis, catachrestical synesthesia	130
Classical metaphors vs. synesthetic metaphors	105, 106, 117, 118
Cratylism	126
Christus Hypercubus	195
Cubism	195
DMT	183
Decadent esthetics	9, 289, 290
Didactic, aphoristic function	274
Ecole Romane	9
Epistemological quality	125
Eutaxy	62
Fourth dimension	194, 195, 197
Heuristic reading vs. hermeneutic reading	210
Hypallage synesthesia	133
Idiolectic pertinence vs. classematic pertinence	111
Futurism	167
Hyperbaton	167

Idiosyncratic quality	118
Impressionism	133
Impressionistic/polyptical synesthesia	133
Isotopy	106, 107, 108
Limbic system	86
Melatonin	183
Metaphor as a semantic bridge or crypt	118, 124
Metaphor as a semantic wormhole	124
Metaphor in absentia vs. metaphor in praesentia	129
Mimesis vs. rhetoricity	126, 127
Minimal pairs	284
Mise en abyme/internal reduplication	165, 166
Musicotherapy	187
Neurological music therapy (NMT)	188
Nouveau Roman	166
Ontogenetic process	56
Phonic mimologism	62
Pineal gland	183
Palingenetic vision of the universe	20, 273
Paratactic structures	273
Poststructuralism and synesthesia en style artiste	289, 290, 291
Psychological rhythm vs. canonical or internal rhythm	68
Semantic indirection	26, 27, 28
Semantic nucleus, matrix	26, 137, 276
Serotonin	183
Symbolist Tree	48
Spiritus Mundi	271, 277
Surrealism	167
Synecdoche/synecdochical synesthesia	131
Synesthesia en style artiste	289, 290, 291
Synesthesiametrics	125

INDEX OF ARTISTS, SCHOLARS, PHILOSOPHERS & AUTHORS

Pages

Aristotle	40, 76, 115
Baron-Cohen	77
Barthes Roland	56, 106, 114, 122
Baudelaire Charles	46, 48, 54, 208, 215
Cytowic Richard	80, 99
Cratylus	76
Cretien Van Campen	170
Dali Salvador	195
Deleuze	166
Derrida	166
Descartes	40
Eco Umberto	107
Einstein Albert	54, 99, 119, 198
Eliot T. S.	27, 160
Feynman Richard	170
Heraclitus	44, 48
Hegel	48
Hugo	156
Kaku Michio	16, 195
Kant Immanuel	54, 55
Locke John	75, 76
Mallarmé Stéphane	155, 161, 164
Parmenides	44, 48
Picasso	154
Plato	76, 151, 202
Pythagoras	60, 76
Ramachandran & Hubbard	102, 109
Rimbaud Arthur	77, 268, 219
Rodenbach Georges	78, 165
Ronsard	147
Swedenborg Immanuel	47, 54, 78
Valéry Paul	160
Verlaine Paul	62
Yeats W. B.	160

www.ingramcontent.com/pod-product-compliance
Lightning Source LLC
Chambersburg PA
CBHW020731180526
45163CB00001B/190